Current Topics in Microbiology and Immunology

158

Editors

Viral Expression Vectors

Edited by N. Muzyczka

With 20 Figures and 5 Tables

Springer-Verlag
Berlin Heidelberg NewYork
London Paris Tokyo
Hong Kong Barcelona
Budapest

NICHOLAS MUZYCZKA Ph.D.
Associate Professor
Department of Microbiology
School of Medicine
State University of New York
at Stony Brook
Stony Brook, N.Y. 11794-8621
USA

ISBN-13:978-3-642-75610-8 e-ISBN-13:978-3-642-75608-5
DOI: 10.1007/978-3-642-75608-5

Softcover reprint of the hardcover 1st edition 1992
Library of Congress Catalog Card Number 15-12910

Typesetting: Thomson Press (India) Ltd, New Delhi

23/3020-543210 – Printed on acid-free paper.

Preface

In the past ten years there has been enormous progress in the development of eukaryotic viral vectors. In general, these vectors have been developed for one of three reasons: to achieve high levels of expression of a particular gene product (poxvirus, baculovirus, and adenovirus), to clone eukaryotic genes in combination with functional assays (Epstein-Barr virus), of for use as delivery vehicles for the stable introduction of foreign genes into mammalian cells (retroviruses, Epstein-Barr virus, and adeno-associated virus). Each vector has its strengths and weaknesses that are rooted in the sometimes bewildering strategies that the parent viruses use for propagation. No one of these vectors is appropriate for all of the problems that a molecular biology laboratory is likely to encounter, and few of us are knowledgeable in the molecular virology of all of these viruses.

This volume represents an attempt by the authors to assemble a review of these vectors in one place and in a form useful to laboratories that do not necessarily have experience with eukaryotic viruses. Clearly, any virus can be modified to serve as a vector for some purposes, and it was not possible to include a description of all of these. In addition, one eukaryotic vector, SV40 (the first one developed), has been reviewed so widely that we saw no reason to include it here. The vectors described in this volume were chosen either because they have already proven their worth in many areas or because we believe they are likely to become widely useful in the near future.

Finally, the success of this volume in fulfilling its purpose is due entirely to the efforts of the authors and to their patience with the editor.

February 1992 NICHOLAS MUZYCZKA

List of Contents

List of Contributors

(Their addresses can be found at the beginning of the respective chapters)

Retroviral Vectors*

A. D. MILLER

1 Introduction

An understanding of the retroviral life cycle and the ability to manipulate the viral genome has allowed the development of gene transfer vectors which utilize the efficient gene transfer capability that retroviruses have evolved. The advantages of currently available retroviral vectors include the ability to efficiently infect a wide range of cell types from different animal species, the

---

Fred Hutchinson Cancer Research Center, 1124 Columbia Street, Seattle, WA 98104, USA
* This work was supported by grants from the National Heart Lung and Blood Institute and the National Cancer Institute of the National Institutes of Health.

Current Topics in Microbiology and Immunology, Vol. 158

precise integration of genetic material carried by the vector into recipient cells, the lack of vector spread or production of viral proteins after infection, and the lack of toxicity of these viruses in infected cells. In this review I will cover features of retrovirus biology relevant to retroviral vector use, different approaches to constructing vectors, and some of the uses for retroviral vectors. Although a great number of individual vectors and virus producing cell lines are currently available, I hope to summarize the important principles to allow their use by those unfamiliar with this gene transfer technique.

2 Retrovirus Life Cycle

2.1 Retroviral Genes and Organization

The structure of a typical retroviral genome is shown in Fig. 1. The virus uses several strategies to make the multiple proteins needed for viral replication (Weiss et al. 1984). Only one promoter, located in the viral long terminal repeat (LTR), is used to transcribe RNA. Splicing of the full-length viral mRNA leads to two RNAs. The full-length mRNA is translated to give the *gag* and *gag-pol* polypeptides and the spliced mRNA is translated to give the *env* polypeptide. The *gag-pol* polypeptide is made by infrequent read-through translation of the stop codon at the end of the *gag* polypeptide. The final step in the production of the individual retroviral proteins is the cleavage of the *gag*, *gag-pol*, and *env* polypeptides into 8–10 mature proteins by a protease encoded in the *pol* gene (the protease can also occur in the *gag* region e.g., in Rous sarcoma virus). The *gag* region encodes the virion core proteins, the *pol* region encodes the protease, reverse transcriptase, and integrase, and the *env* region encodes the viral surface proteins, or envelope proteins.

2.2 Life Cycle

The steps in the retroviral life cycle are depicted in Fig. 2. Virus entry into cells is dependent on the presence of receptors which bind the viral envelope protein. After virus fusion with the cell membrane, the single strand viral RNA is converted to double strand DNA by reverse transcriptase. The viral RNA and DNA remain associated with the intact viral core during this process and presumably until integration of the viral DNA into the genomic DNA of the infected cell (P.O. Brown et al. 1987). Integration of the viral DNA is precise with respect to the retrovirus genome and is mediated by sequences at the ends of the LTRs. Integration is semirandom with respect to the host cell genome in that many sites are available for integration, but some sites are used much more frequently (Shih et al. 1988). RNA is transcribed from the integrated

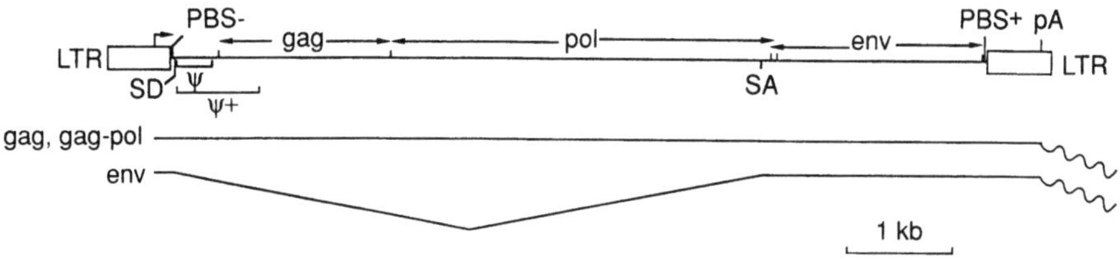

Fig. 1. Structure of the proviral form of a typical retrovirus (MoMLV) and pattern of transcription products. *PBS-*, primer binding site for minus strand DNA synthesis; *PBS +*, plus strand primer binding site; *SD*, splice donor; *SA*, splice acceptor; *LTR*, long terminal repeat; ψ, packaging signal; ψ^+, extended packaging signal; *pA*, polyadenylation signal; *gag*, *pol* and *env*, viral protein coding regions

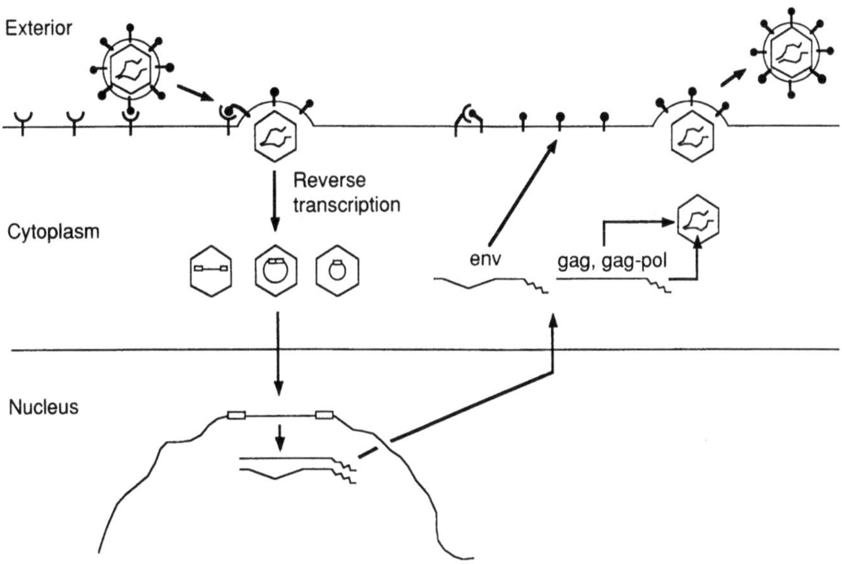

Fig. 2. Retrovirus life cycle

provirus and is translated to make the viral proteins, which combine with the full-length viral RNA to make new virions. The full-length viral RNA contains signals which direct specific packaging of viral RNA into virions, termed ψ or E (for encapsidation) sequences (LINIAL and MILLER 1990).

2.3 Host Range

Host range originally referred to the range of animal species which would serve as hosts for replication of a given retrovirus. Here we will define the host range of a retrovirus as the range of cell types as well as the range different species that can be infected with and will express a given retrovirus, this definition being more relevant for gene transfer studies. Host range is primarily determined by the viral envelope protein, which determines the ability of a virus to bind and enter cells. By far the most useful host range for gene transfer studies is afforded by the envelope from amphotropic murine leukemia viruses, which allows infection of many types of cells from a wide range of vertebrate species (HARTLEY and ROWE 1976; RASHEED et al. 1976). Another important factor in host range is the promoter and enhancers carried by the virus, which will affect the ability of the virus to express in a given cell type (WAGNER et al. 1985; FRANZ et al. 1986; HOLLAND et al. 1987; LINNEY et al. 1987; TAKETO and TANAKA 1987; HILBERG et al. 1987).

3 Oncogenic Retroviruses as Vectors

The ability of acutely oncogenic retroviruses to induce uncontrolled cell growth is the result of viral acquisition of host genetic material. As such, these viruses serve as naturally occurring examples of retroviral vectors and indicate possible strategies for vector construction. A survey of oncogenic retroviruses (VAN BEVEREN and VERMA 1986) reveals that in most cases the acquired coding regions are fused in frame with existing retroviral coding regions so that the resultant transforming proteins are fusion proteins. Exceptions include Rous sarcoma virus (RSV), in which the *src* oncogene is expressed from a subgenomic spliced mRNA, and FBJ-MSV, in which the transforming protein *fos* is not a fusion protein but is translated beginning with the normal *fos* initiation codon. Expression of all of these transforming proteins is driven by the promoter and enhancers located in the viral LTR; none have acquired new promoters for expression of the transforming gene. This phenomenon led to early speculation that internal promoters might not work well is retroviral vectors (EMERMAN and TEMIN 1984b). However, the lack of internal promoters in acutely oncogenic viruses may be due more to the events leading to acquisition of the new sequences, which may involve recombination at the RNA level, than to inherent limitations of this strategy for gene expression.

4 Replication-Competent Retroviral Vectors

Rous sarcoma virus carries the *src* oncogene in addition to a full complement of genes required for replication, and thus provides the earliest example of a replication-competent retroviral vector. RSV has been exploited as a vector by replacement of the *src* gene with other cDNAs. Early attempts to use RSV as a vector were hindered by the presence of direct repeats at either end of the *src* gene which led to frequent deletion of new inserts, but these problems have been solved in currently available vectors (HUGHES et al. 1987). Vectors derived from RSV efficiently infect avian cells, and can infect mammalian cells but with reduced efficiency. Replication-competent vectors based on Moloney murine leukemia virus (MoMLV) have also been described which include a tRNA suppressor gene (LOBEL et al. 1985; REIK et al. 1985) or a mutant dihydrofolate reductase (dhfr*) cDNA (Stuhlmann et al. 1989). These vectors allow efficient infection of mammalian cells and thus enlarge the applicability of replication-competent vectors.

5 Replication-Defective Vectors

The majority of applications of retroviral vectors require that the vector not spread after initial cell infection. This can be accomplished by replacement of most or all of the coding regions of a retrovirus with the gene(s) to be transferred, so that the vector by itself is incapable of making proteins required for retrovirus replication. Viral proteins needed for the initial round of vector infection can be provided by retrovirus packaging cells (sec Sect. 6.1), resulting in production of virus which can infect cells but which cannot replicate further.

5.1 Methods for Expressing Genes

Three general strategies have been used to provide for gene transcription in retroviral vectors. These are linkage of the gene to the LTR promoter, the use of alternative splicing to express one or more genes from the LTR promoter, or linkage of gene(s) to internal promoter(s) (Fig. 3). Although a study of gene insertion in naturally occurring oncogenic viruses (see Sect. 3) might suggest that fusion proteins created between viral and inserted genes might be the best method for expression of exogenous genes, this approach has not been widely used except in the study of preexisting fusion products in oncogenic viruses. In addition, although oncogenic viruses and other naturally occurring retroviruses have not been found to incorporate internal promoters, many examples now exist of retroviral vectors which faithfully transmit and express genes by using internal promoters.

Vectors containing one gene often contain that gene linked to the viral LTR (Fig. 3a). The presence of the retroviral packaging signal between the viral promoter and the translational start codon of the inserted gene appears not to affect gene expression, even though several ATG start codons are present in the packaging signal of commonly used vectors. This construction strategy works well when the gene of interest is also a selectable marker, as in the case of hypoxanthine-guanine phosphoribosyl transferase (hprt) (MILLER et al. 1983).

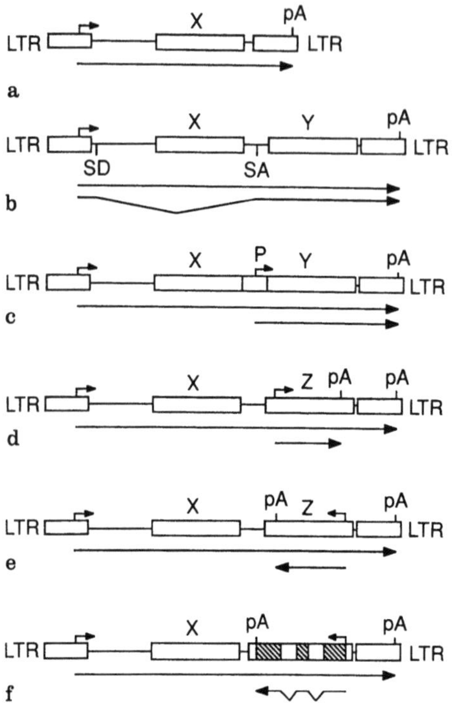

Fig. 3 a–f. Typical retroviral vectors. *LTR* indicates the retroviral long terminal repeats, *X* and *Y* indicate protein encoding regions, *Z* indicates a minigene (gene without introns), *pA* indicates a polyadenylation signal, *arrows above* the vector diagrams indicate promoters and direction of transcription, *arrows below* the vector diagrams indicate RNA species transcribed from the vectors. Vector construct **f** contains an intact gene with introns, and the coding regions of the gene are cross-hatched. The internal *pA* site in constructs **e** and **f** provides for polyadenylation of RNAs transcribed in reverse orientation with respect to virus transcription, while other *pA* sites function in the direction of virus transcription

Of the vectors which employ alternative splicing to express genes from the viral LTR, the SV(X) vector (CEPKO et al. 1984) has been widely used to express a desired gene and the neomycin phosphotransferase gene as a selectable marker (Fig. 3b). This type of vector is modeled on the structure of the parental helper virus (MoMLV) in which the *gag* and *gag-pol* proteins are translated from the full-length viral mRNA while the *env* protein is made from a spliced mRNA (Fig. 1). Thus, one of the proteins encoded by the vector is translated from the full length vector mRNA, while partial splicing from a splice donor near the 5′ LTR to a splice acceptor just upstream of the second gene yields a transcript which encodes the second gene product. One drawback of this strategy is that foreign sequences are inserted into the intron of the spliced gene. This can affect the ratio of spliced to unspliced mRNAs or, at worst, provide alternative splice acceptors which preclude production of the spliced mRNA encoding the second gene product (KORMAN et al. 1987). These effects are unpredictable and affect the production of the two encoded proteins.

Vectors containing internal promoters have been widely used to express multiple genes as well as to exploit promoter/enhancer combinations other than the viral LTR for driving gene expression (Fig. 3c–f). One such virus included two different internal promoters which, together with the viral LTR, were used to transcribe three different cDNAs (OVERELL et al. 1988). A potential problem with this strategy was indicated by experiments involving vectors carrying two selectable markers driven by different promoters. In some cases selection for one of the genes appeared to result in reduced expression from the other gene (EMERMAN and TEMIN 1984a). This effect was called promoter suppression. Thus, for a vector carrying a selectable marker and a gene of interest, selection for cells containing the vector might result in poor expression from the gene of interest. However, this effect is highly dependent on vector construction (EMERMAN and TEMIN 1986a, b) and at least in some cases is not observed (PALMER et al. 1987). In addition, because expression of the "suppressed" gene was always compared with expression of the same gene in cells selected for expression of the gene (and thus which express the gene at a high level in order to survive selection), it is difficult to tell whether low expression can really be attributed to promoter suppression or is simply an artifact due to the way low expression was defined.

Because retroviral replication involves an RNA intermediate, all strategies for inserting genes into a retrovirus must permit full-length transcription of the vector genome. For example, insertion of a polyadenylation signal should be avoided since premature polyadenylation within the vector will reduce if not prevent full-length vector transcription and the production of recombinant virus (SHIMOTOHNO and TEMIN 1981; MILLER et al. 1983). In addition, introns contained within the insert may be removed during vector replication (SHIMOTOHNO and TEMIN 1982; SORGE and HUGHES 1982; KRIEGLER et al. 1984; CEPKO et al. 1984). Vectors which rely on alternative splicing for gene expression (Fig. 3b) retain their introns because the packaging signal is contained within the intron in these vectors, and RNAs without this signal are not efficiently encapsidated and replicated (MANN and BALTIMORE 1985).

5.2 Selectable Markers

A variety of selectable markers have been successfully included in retroviral vectors. These include the bacterial neomycin and hygromycin phosphotransferase genes that confer resistance to G418 and hygromycin respectively (PALMER et al. 1987; YANG et al. 1987), a mutant mouse dihydrofolate reductase gene (*dhfr**) which confers resistance to methotrexate (MILLER et al. 1985), the bacterial *gpt* gene which allows cells to grow in medium containing mycophenolic acid, xanthine, and aminopterin (MANN et al. 1983), the bacterial *hisD* gene which allows cells to grow in medium without histidine but containing histidinol (DANOS and MULLIGAN 1988), and the multidrug resistant gene (*mdr*) which confers resistance to a variety of drugs (PASTAN et al. 1988; GUILD et al. 1988). All of these markers are dominant selectable markers and allow chemical selection of most cells expressing these genes. An exception is the *dhfr** gene which apparently does not provide a selective advantage in human diploid fibroblasts because these cells are resistant to the effects of methotrexate (PALMER et al. 1987). β-Galactosidase provides a dominant marker in that cells expressing this gene can be selected by using the fluorescence-activated cell sorter (NOLAN et al. 1988). In fact, any cell surface protein can provide a selectable marker for the presence of a vector in cells not already making the protein. Cells expressing the protein can be selected by using a fluorescent antibody to the protein and a cell sorter. Other selectable markers that have been included in vectors include the *hprt* (MILLER et al. 1983; CHANG et al. 1987) and thymidine kinase (*tk*) (SHIMOTOHNO and TEMIN 1981; WEI et al. 1981) genes which allow the growth of cells in medium containing hypoxanthine, amethopterin, and thymidine. For these selections to work, the infected cells must be initially *hprt* or *tk* deficient respectively.

5.3 Types of Gene Inserts

The most widely used and most easily accommodated insert is an intronless cDNA copy of an mRNA encoding a given protein. Even large cDNAs with repetitive internal sequences, such as those found in a collagen cDNA, can be faithfully transmitted (STACEY et al. 1987). One might have predicted that the ability of reverse transcriptase to switch templates during virus replication would have led to deletions of such repetitive sequences, but at least in this case, deletions were not observed. The 5′ and 3′ noncoding regions of the cDNA can be trimmed to reduce the size of the insert and to remove potential polyadenylation signals that may occur in the 3′ end of cDNAs (SHIMOTOHNO and TEMIN 1981; MILLER et al. 1983). The cDNA is generally inserted in the same transcriptional orientation as the viral LTR, and can be expressed by any of the strategies described in Sect. 5.1.

Retrovirus vectors have also been used to make antisense RNAs by reversing the orientation of the cDNA with respect to a promoter. Antisense RNAs have been used to inhibit infection of cells by retroviruses containing the sense strand RNA (TO et al. 1986) and to inhibit expression of proteins from sense strand RNAs (TREVOR et al. 1987).

Other inserts which have been used in retroviral vectors include intronless "minigenes" Fig. 3D and E) which are normal genes from which the introns have been removed (MILLER et al. 1984; BENDER et al. 1988). Minigenes contain the normal promoter, upstream enhancers, and transcription termination signal present in the parental gene and are thus potentially responsive to transcriptional factors that normally control gene activity. The introns are removed to reduce the size of the insert and to avoid interference with the retroviral life cycle. Since these inserts contain the normal polyadenylation signal for the gene, one might expect that minigenes must be inserted into vectors in the reverse orientation. However, several retroviral vectors containing forward orientation minigenes permit faithful gene transfer without substantial loss in virus titer (less than five fold) (MILLER et al. 1984; BENDER et al. 1988), suggesting that there is enough transcription through at least some polyadenylation signals to allow production of full-length viral RNA.

Entire genes containing introns have also been used in retroviral vectors (Fig. 3f) (BANDYOPADHYAY and TEMIN 1984; CONE et al. 1987a, b; KARLSSON et al. 1987; STOECKERT et al. 1987; BENDER et al. 1988; DZIERZAK et al. 1988; MILLER et al. 1988; BENDER et al. 1989). In this case the gene must be inserted in reverse orientation to prevent removal of the introns during vector replication. This strategy is particularly important for genes containing regulatory elements within their introns, such as the immunoglobulin enhancer (CONE et al. 1987a) or elements that map to the β-globin second intron which are required for high-level gene expression (MILLER et al. 1988). Reverse-orientation gene inserts in general have deleterious effects on vector titer, which can be ameliorated by vector modification (MILLER et al. 1988). Such vectors can still yield virus at titers high enough to permit infection of poorly infectable cells, such as hematopoietic stem cells (DZIERZAK et al. 1988; BENDER et al. 1989).

5.4 Other Useful Inserts

A variety of sequences have been included in retroviral vectors to extend their utility, especially for genetic studies. A bacterial plasmid origin of replication and drug resistance genes which function in bacteria (*amp* and *neo*) have been included to facilitate the cloning of vector and flanking sequences from infected cells (CEPKO et al. 1984; BERGER and BERNSTEIN 1985; JHAPPAN et al. 1986). Similarly, tRNA suppressor genes can be included in retroviral vectors to allow subsequent cloning of proviral and flanking cellular sequences in bacteriophage libraries (LOBEL et al. 1985; REIK et al. 1985). Inclusion of the SV40 origin of replication in a vector allows amplification of vector and flanking sequences by fusion of vector infected cells to COS cells. COS cells constitutively express SV40 T antigen, which induces multiple rounds of replication at the SV40 origin carried by the vector and the production of circular SV40 replicons which have been formed by recombination between vector sequences and/or flanking cellular sequences. The low molecular weight DNA obtained by Hirt extraction of such fused cells is then a highly enriched source of vector DNA and the sequences flanking the vector integration site. In combination these techniques provide simple methods for rescuing and recloning recombinant proviruses and flanking cellular sequences.

A highly innovative approach to the study of immunoglobulin gene rearrangement was devised by using retroviral vectors that contain sequences from the immunoglobulin locus that are involved in immunoglobulin rearrangement (LEWIS et al. 1984; LANDAU et al. 1987; DESIDERIO and WOLFF 1988). Although the same studies can be done by simply transfecting similar immunoglobulin sequences into cells, the advantage of using retroviral vectors is that the viral sequences are inserted in a precise, predictable manner into recipient cells, and conditions for integration of a single copy per cell are easily achieved. The vectors were designed such that proper rearrangement of these sequences within the vector would result in expression of a drug resistance gene. Thus, cells containing a rearranged vector could be identified by drug selection. Cloning of the rearranged immunoglobulin genes permitted the subsequent sequencing of the rearranged joints to provide a fine structure analysis of the process of rearrangement.

5.5 Design of High-Titer Vectors

Helper virus-free vector titers of just over 10^7 cfu/ml are obtainable with current vector designs (see Sect. 6.1). Often experiments can be done with vectors produced at much lower titers, but for practical reasons high titer virus is desirable, especially when a large number of cells must be infected. In addition, high titers are a requirement for infecting certain difficult to infect cells. For example, the frequency of human hematopoietic progenitor cell infection is strongly dependent on vector titer and useful frequencies of infection only occur with very high titer viruses (Hock and Miller 1986; Hogge et al. 1987).

Comparison of different vector designs allowed the definition of essential elements for high-titer virus production. Early vector designs showed that almost all of the internal protein-encoding regions of murine leukemia viruses could be deleted while still retaining a transmissible vector (Miller et al. 1983). These early vectors retained only a small portion of the 3′ end of the *env* coding region. Subsequent work has shown that all of the *env* gene coding sequences can be removed without further reduction in vector titer (J. Morgenstern and H. Land, personal communication). Only the viral LTRs and short regions adjoining the LTRs, including plus and minus strand primer binding sites and a region required for packaging of viral RNA into virions (the ψ site, Mann et al. 1983) were needed for vector transmission. However, virus titers obtained with these early vectors were still about ten fold lower than the parental helper virus. Later work showed that retention of sequences at the 5′ end of the *gag* gene significantly raised virus titers, and that this effect was due to more efficient packaging of viral RNA into virions (Armentano et al. 1987; Bender et al. 1987; Adam and Miller 1988). This effect was not due to viral protein synthesis from the *gag* region of the vector because disruption of the *gag* reading frame or alteration of the *gag* start codon to a TAG stop codon had no effect on vector titer (Bender et al. 1987). These recent experiments have led to the conclusion that the retroviral packaging signal in murine leukemia viruses is actually larger than the ψ signal previously defined by deletion analysis (Mann et al. 1983). Thus, it is important that this larger signal, called ψ^+ (Fig. 1), be included in retroviral vectors to obtain high titers, which range from 10^6 to over 10^7.

5.6 Self-Inactivating Vectors

Several groups have designed vectors in which the enhancers or the enhancers and promoter in the 3′ LTR are deleted (Yu et al. 1986; Hawley et al. 1987; Yee et al. 1987; Dougherty and Temin 1987). After one round of vector replication, this deletion is copied to the 5′ LTR as well as the 3′ LTR, producing an inactive provirus. However, promoters located internal to the LTRs in such vectors will still be active. This strategy has been employed to eliminate effects of the enhancers and promoters in the viral LTRs on transcription from internally placed genes, which can lead to enhancement (Jolly et al. 1983) or repression (Emerman and Temin 1984a) of expression from the internal gene. Another application of this strategy is to prevent possible downstream transcription from the 3′ LTR into genomic DNA (Herman and Coffin 1987). This is of particular concern in the case of preventing possible oncogene activation in the context of the potential application of these vectors to treat human genetic disease, or gene therapy. Drawbacks of this strategy include the lower titer of self-inactivating vectors in comparison with vectors having intact LTRs (at least ten fold lower), and rearrangement of current vectors to forms having intact LTRs, presumably by recombination of the vector with itself or with viral sequences in the retrovirus packaging cells used to produce virus.

6 Production of Virus from Vector Constructions

Unlike bacteriophage assembly which can be accomplished in a cell-free system, production of retrovirus virions has only been accomplished in intact cells. To make virus from the DNA form of a replication-competent retrovirus vector, the DNA is introduced into suitable cells (cells in which the virus can replicate to high titer) by any of a variety of physical techniques and virus production follows. It is not necessary to introduce the DNA into every cell because the virus will spread to infect all cells. To make virus from replication-defective retrovirus vectors the DNA form of the vector is either introduced into cells along with replication-competent "helper" virus, in which case the resultant virus is a mixture of vector and helper virus, or the vector is introduced into packaging cells which results in the production of helper-free vector. Packaging cells are designed to provide all viral proteins but not to package or transmit the RNAs encoding these functions. Thus retroviral vectors produced by using packaging cells can infect cells but cannot replicate further.

6.1 Retrovirus Packaging Cell Lines

A major determinant in choosing a packaging cell line is that the virus obtained has the proper host range to allow infection of the desired cells. Host range is

Table 1. Retrovirus packaging cell lines

Type[a]	Name	Host range	Maximum titer[b]	Reference
A	ψ-2	Ecotropic	10^7	MANN et al. 1983
	ψ-AM	Amphotropic	2×10^5	CONE and MULLIGAN 1984
	PA12	Amphotropic	10^7	MILLER et al. 1985
	Q4dh	Avian	3×10^5	STOKER and BISSELL 1988
B	C3A2	Avian, rat, dog	10^7	WATANABE and TEMIN 1983
	Clone 32	Ecotropic	3×10^4	BOSSELMAN et al. 1987
C	T19-14X	Amphotropic	10^3	SORGE et al. 1984
	PA317	Amphotropic	10^7	MILLER and BUTTIMORE 1986
D	ψCRE	Ecotropic	10^6	DANOS and MULLIGAN 1988
	ψCRIP	Amphotropic	10^6	DANOS and MULLIGAN 1988
E	GP + E-86	Ecotropic	4×10^6	MARKOWITZ et al. 1988

[a] Packaging cell type based on type of deleted helper virus (Fig. 4)
[b] Highest reported titers; in some cases this value is from papers published after the initial report describing the cell line

dependent on the parental virus used to construct the packaging cells. There are many complicated viral determinants that modify the ability of a given virus to infect different cell types from different species (WEISS et al. 1984), the most important of which is the viral envelope protein. Packaging cell lines have been developed by using coding regions from a variety of different retroviruses, including avian leukosis virus (STOKER and BISSELL 1988), reticuloendotheliosis virus (WATANABE and TEMIN 1983), and ecotropic and amphotropic murine leukemia viruses (Table 1).

The most useful host range is provided by amphotropic packaging cells, which produce virus capable of infecting human, mouse, rat, cat, mink, monkey, dog, and chicken cells (MILLER and BUTTIMORE 1986). Notable exceptions to this list are hamster and bovine cells, which can not be efficiently infected with virus from amphotropic packaging cells. Other widely used packaging cells are based on an ecotropic murine leukemia virus, MoMLV (MANN et al. 1983). Ecotropic viruses are capable of infecting rodent cells (e.g., mouse and rat) but do not have a wide host range and cannot infect human, cat, monkey, or dog cells. Packaging cells based on reticuloendotheliosis virus can infect avian cells as well as dog and rat cells. Packaging cell lines based on viruses with alternate host ranges are constantly being developed to enlarge the range of infectable cell types.

Other important considerations in choosing a packaging cell line are the titer of virus produced and the potential of the packaging cells to produce helper virus. Several reported packaging cell lines are capable of producing vectors at titers of up to 10^7 cfu/ml in the absence of helper virus (Table 1). The tendency of a packaging cell line to produce helper virus is dependent on the degree to which the helper virus used to make the cells is disabled and on the particular vector used (Fig. 4). Packaging cell lines of type a or b require only one

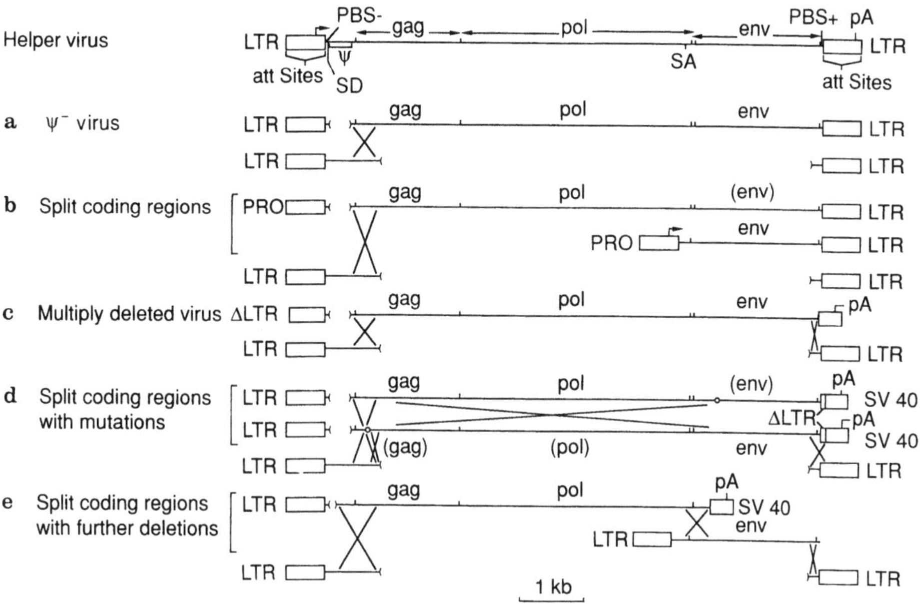

Fig. 4 a–e. Strategies for packaging cell line construction. A generic helper virus (modeled on MoMLV) is shown at the *top*. Strategies used to generate deleted viruses that provide all proteins required for viral replication but which cannot themselves replicate are shown. Deleted viruses of types **a** and **c** involve single DNA constructs, while those of types **b**, **d** and **e** involve two constructs. The *bottom construct* in each set shows a generic retroviral vector (or other endogenous retrovirus-like element present in packaging cells). Potential mechanisms for homologous recombination leading to helper virus production are indicated by *crosses*. *PRO*, promoter (LTR, WATANABE and TEMIN 1983, or metallothionine promoter, BOSSELMAN et al. 1987) -o-, mutation; *ΔLTR*, deleted LTR; *pA*, polyadenylation signal (SV40, MILLER and BUTTIMORE 1986 or truncated LTR, SORGE et al. 1984) *SD*, splice donor; *SA*, splice acceptor

recombinational event between the retroviral vector and the defective helper virus in the packaging cells to produce helper virus. Packaging cells of type c require two events to produce helper, and have been shown to be much less prone to produce helper virus than packaging cells of type a (MILLER and BUTTIMORE 1986). Packaging cells of type d require two or three recombinational events to generate helper virus depending on the position of the *gag* mutation and the amount of *gag* sequences in the retroviral vector. Three recombinational events are required to generate helper virus from packaging cells of type e, thus these cells are presumably the least likely to generate helper virus.

Another approach to minimizing the possibility of helper virus production is to modify the retroviral vector to avoid homologous overlap with the deleted helper virus. While overlap at the 5′ end may be difficult to avoid due to the presence of elements in this area needed for high titer vector production (see Sect. 5.5), overlap at the 3′ end can easily be avoided without a reduction in vector titer (J. MORGENSTERN and H. LAND, personal communication).

Even with improved packaging cell lines, helper virus production is still possible. Thus, much time can be saved in the long run by screening for helper virus in experiments where the lack of viral spread is critical. The best assay for helper virus is probably a functional assay involving marker rescue. Tester cell lines carrying a selectable retroviral provirus are infected with a large amount of the virus to be tested, passaged for 2 weeks to allow spread of possible helper virus throughout the population, and the medium from the infected cells is tested for the presence of virus conferring drug resistance to naive cells. If transfer of drug resistance is detected, the selectable provirus was rescued because replication-competent virus was present in the initial virus stock. It is important to be sure that the tester cells allow replication of likely helper virus contaminants. For example, human tester cells cannot replicate ecotropic viruses and thus cannot be used to test for these viruses.

Two methods have been used to introduce vectors into packaging cells, by direct transfection of DNA into the cells or by infection with virus produced from other packaging cells (Miller et al. 1986). Whereas the first method can result in insertion of multiple copies of the vector into the recipient cells and to the production of rearranged virus, the latter method results in single copy integrated proviruses which can be screened for the proper structure and which generally yield unrearranged virus (Bender et al. 1988). One might expect that direct transfection of vector sequences might lead to higher virus titers because of the presence of multiple copies of the vector, however, the titer of virus from infected cells containing single proviruses is at least as high if not higher. The origin of replication and T antigen coding region from polyoma virus have been incorporated into plasmid sequences in an attempt to increase virus titer by generating multiple copies of the plasmid in transfected cells, but again the titers are no higher than those of packaging cells containing single integrated proviruses (Korman et al. 1987).

Once a packaging cell line has been isolated that contains an unrearranged vector and produces the vector at high titer in the absence of helper virus, this line provides a continuous source of the vector. The cells can be frozen by using standard techniques and thawed for use at any time. Infection of recipient cells is performed by exposure of the cells to virus-containing medium harvested from the packaging cells, or in the case of cells that are difficult to infect, by direct cocultivation with vector-producing cells. The disadvantage of the cocultivation technique is that the infected cells must be separated from the packaging cells after infection. Virus-containing medium to be used for infecting cells is usually produced by feeding a just confluent dish of vector producing cells with fresh culture medium and harvesting the medium after 12–24 h. Virus-containing medium can be frozen for years at $-70°$ C without loss of virus titer, but thawing and refreezing will decrease the titer.

We have noted some instability of packaging function in packaging cell lines (Bender et al. 1987). For example, after growth of PA317 cells for 4 months in culture it was noted that after infection of these cells with a retroviral vector, only 20% of the clones containing the vector produced high-titer virus. In addition, many of the cells did not express the HSV TK gene that was used for cotransfection of the helper virus DNA that encodes the viral proteins. Selection for the TK^+ phenotype by using HAT selection restored packaging function in the majority of cells, i.e., infection of these cells with a retroviral vector resulted in a high percentage of clones which secreted high-titer virus. Thus, it is important to avoid prolonged passage of packaging cells and vector-producing packaging cells by freezing and using early passage stocks of cells. This practice will also minimize the possibility of helper-virus production, which is more likely following longer periods of cell culture.

7 Infectable Cells

Most immortal cultured mammalian cells can be infected with amphotropic retroviral vectors. Certain hematopoietic cell lines such as murine erythroleukemia cells and HL-60 promyelocytic leukemia cells are poorly infectable, but still can be infected at rates 10- to 1000-fold lower than those of mouse fibroblasts (Bender et al. 1987; Collins 1988). Hamster and bovine cells are very poorly infectable apparently due to a lack of receptors for the viral envelope protein.

Cells with finite life span in culture can also be infected. Human skin fibroblasts can be infected at about 50% efficiency (Palmer et al. 1987) and human keratinocytes at up to 0.5% efficiency (Morgan et al. 1987). Primary hepatocytes can be infected at efficiencies of up to 25% (Wilson et al. 1988a). Many cell types capable of further differentiation, such as mouse, dog, and human hematopoietic progenitor cells, can be infected at efficiencies of from 5%–20% (Dick et al. 1985; Hock and Miller 1986; Kwok et al. 1986). Pluripotent hematopoietic stem cells from mice can be infected with greater than 10% efficiency (Dick et al. 1985; Keller et al. 1985; Lemischka et al. 1986; Bowtell et al. 1987). Infection of these cells has been demonstrated by showing that particular unique viral integrants can be detected in all myeloid and lymphoid cell lineages in a mouse receiving infected bone marrow cells, and by showing that these integrants can be detected over long periods of time in vivo. The percentage of transplanted cells that contain a virus carrying the *neo* gene can be improved substantially by preselecting the bone marrow cells in G418 for 24–48 h before transplanting the cells into animals to kill cells that do not express the vector (Dick et al. 1985; Keller et al. 1985).

Retroviral vectors can also be used to infect totipotent cells in mouse and chicken embryos at useful frequencies, and animal strains carrying integrated proviruses have been obtained (Van der Putten et al. 1985; Jahner et al. 1985; Huszar et al. 1985; Soriano et al. 1986; Soriano and Jaenisch 1986; Stewart et al. 1987; Salter et al. 1987). Mouse embryonic stem cells can also be infected at high efficiency with retroviral vectors, and cell lines carrying at least 15 proviruses can be generated by cocultivation of embryonic stem (ES) cells with virus-producing cells (ROBERTSON et al. 1986). Since ES cells can contribute to the germ line, strains of mice carrying many semirandomly integrated proviruses (SHIH et al. 1988) can be generated. Such animals should provide a rich source of interesting mutations, each of which is marked by the presence of the integrated provirus.

Retroviral vectors have been used to infect replicating neuronal cells in vivo (PRICE et al. 1987; TURNER and CEPKO 1987). After infection, the provirus provides a marker for all progeny cells and thus the fates of neuronal cells can be followed. Incorporation of the β-galactosidase gene in these vectors allows identification of infected cells by a chemical reaction which stains infected cells blue.

8 Uses for Retroviral Vectors

In addition to the general use of retroviral vectors to transfer and express genes, several other applications are worth noting. Because vector integration does not generally lead to gross rearrangements of the genome of the infected cell, and because the vector does not mobilize after the initial infection, retroviral vectors have been used as mutagens. For example, *hprt*-deficient mice have been generated from ES cells infected with a retroviral vector which inactivated the *hprt* locus (KUEHN et al. 1987). After mutagenesis, the presence of the vector at the site of the mutation provides a tag to mark the mutated gene. Thus, this technique provides a method to clone genes that have a selectable phenotype after mutagenesis. This technique has been shown to work in a model system involving mutations in the *hprt* gene (KING et al. 1985). These mutagenesis techniques are limited to haploid genes or genes where mutation of one of two copies of the gene results in a selectable phenotype. Since the frequency of retroviral integrations into an individual cell can be controlled by varying the amount of virus used or the time of infection, many mutational events can be produced in a single cell.

Retrovirus vectors have been used as markers for chromosomes in cell fusion studies (NELSON et al. 1984; LUGO et al. 1987). Any chromosome can be marked by using this technique. Further, any location on a given chromosome can be marked, providing a method to select for small regions of a chromosome (WEIS et al. 1984). Again, the ability to control the number of vector integrants in an infected cell gives this technique wide flexibility.

Retroviral vectors have been used to generate cDNAs from genomic copies of genes (KRIEGLER et al. 1984; KAPLAN et al. 1987; AUFFRAY et al. 1987; NODA et al. 1986; M. BROWN et al. 1986; JAT et al. 1986; FELDMAN et al. 1987; MORGAN et al. 1988). Additional elements included in the vector then allow facile recloning of the vector after replication (see Sect. 5.4). This technique is particularly useful for isolating alternate cDNAs which arise from differential gene splicing, such as the SV40 large and small T antigens (KRIEGLER et al. 1984).

Retroviral vectors have potential utility as vectors for cDNA libraries, and have been used in a model system to clone a rat thymidine kinase cDNA (MURPHY and EFSTRATIADIS 1987). The technique entails construction of a cDNA library in a retroviral vector, followed by transfection of the DNA into retrovirus packaging cells to generate virus carrying the inserted cDNAs. Cells are then infected with the resultant virus and screened for expression of the desired cDNA. The advantage of using retrovirus vectors in this application is that cells which are difficult to transfect could be used as recipients for screening cDNA expression libraries.

Retroviral vectors are being considered for the genetic modification of human somatic cells in order to correct genetic disease, or gene therapy (ANDERSON 1984). As currently envisioned, gene therapy would involve removal of some somatic cells, infection of those cells in vitro, and reintroduction of the cells into the patient. Several cellular targets are being considered, including hematopoietic (HOCK and MILLER 1986; KANTOFF et al. 1987; STEAD et al. 1988; DZIERZAK et al. 1988; KARLSSON et al. 1988; BENDER et al. 1989), skin (PALMER et al. 1987; GARVER et al. 1987; ANSON et al. 1987; ST LOUIS and VERMA 1988; OSBORNE and MILLER 1988; PALMER et al. 1989), and hepatic cells (WOLFF et al. 1987; LEDLEY et al. 1987; WILSON et al. 1988b).

9 Limitations

A limitation of retroviral vectors is that inserted sequences must be compatible with the retroviral life cycle. As described above, inserted sequences must allow full-length transcription of the retrovirus genome. In addition, there are limits on the overall size of retroviral vectors. The genome size of a typical replication-competent murine retroviruses is about 8.3 kb while that of RSV, which contains *src* sequences in addition to the normal complement of viral replication genes, is about 9.2 kb. The maximum size for a replication-competent vector based on spleen necrosis virus was similar, about 10 kb (GELINAS and TEMIN 1986). While this size limit precludes insertion of large genes, most cDNAs can be accommodated along with a selectable marker.

Genomic rearrangements and mutations can occur during the generation of retroviral vector stocks. Both problems can be minimized by harvesting virus from packaging cells that contain a single integrated virus with the correct structure. The mutation rate for a single round of virus replication has been estimated to be 2×10^{-5} per base pair per replication cycle (DOUGHERTY and TEMIN 1988).

Various contaminants of retroviral vector stocks may limit their usefulness. Helper virus production can be controlled by using currently available packaging cell lines, but will always be a possibility and should be monitored. More insidious are other endogenous mRNAs in mammalian cells which can be packaged into virions and integrated into recipient cells. For example, VL30 sequences, which are transcribed from retrovirus-like elements present in mouse and rat cells, can be transmitted following infection of these cells with replication-competent helper virus (SHERWIN et al. 1978; BESMER et al. 1979; SCOLNICK et al. 1979). Packaging cells can also package VL30 sequences into infectious virions (RODLAND et al. 1987). These elements often do not contain significant open reading frames for protein translation, but the effects of their transfer are not known.

10 Safety Issues

Because amphotropic retroviral vectors can infect human cells, the question arises of the safety of investigators handling these materials. While there can be no guarantee that these materials are safe, several arguments suggest that the dangers are minimal. Because retroviruses are surrounded by a lipid membrane derived from the virus-producing cell, they are sensitive to desiccation and are readily inactivated by detergents or ethanol. Second, human complement is known to inactivate murine retroviruses, including the amphotropic variety, by direct lysis of virions in an antibody-independent manner (WELSH et al. 1975, 1976; COOPER et al. 1976). This is in contrast to the inability of human serum to inactivate retroviruses which cause disease in humans, HTLV-1 (HOSHINO et al. 1984) and HIV (BANAPOUR et al. 1986). The site of complement action is the viral envelope protein (BARTHOLOMEW et al. 1978). Recombinant viruses between MoMLV and amphotropic virus 4070A, upon which most amphotropic packaging cell lines are based, are also sensitive to complement inactivation R.A. HOCK and A.D. MILLER, unpublished results). This activity should reduce the possibility of viral infection or virus spread *in vivo*. Third, retroviral vectors are generally made in the absence of helper virus, thus preventing their further spread after initial infection. Presumably there would be little consequence of infection of a few somatic cells with a retroviral vector encoding a selectable marker or a normal cellular gene. Oncogenic retroviruses present a different concern in that theoretically one cell infected with such a virus might lead to neoplastic growth of that cell, thus, more care should be exercised with oncogenic retroviruses having an amphotropic coat to prevent contact with virus-containing medium or production of virus-containing aerosols.

11 Conclusion

The technology of retroviral vector design and use has reached a point of maturity where it can be used as a standard tool for gene transfer. While improvements in vector design will continue to be made, the majority of current research papers employ retroviral vectors to address other questions. Retroviral vectors are the gene transfer technique of choice for high-efficiency gene transfer, and their precise integration lends itself to genetic studies. Their potential application to human gene therapy is another promising application which in the near future may lead to a radically new approach to the treatment of disease.

Acknowledgements. I thank Maxine Linial and John Dick for critiques of this manuscript.

References

Adam MA, Miller AD (1988) Identification of a signal in a murine retrovirus that is sufficient for packaging of nonretroviral RNA into virions. J Virol 62: 3802–3806

Anderson WF (1984) Prospects for human gene therapy. Science 226: 401–409

Anson DS, Hock RA, Austen D, Smith KJ, Brownlee GG, Verma IM, Miller AD (1987) Towards gene therapy for hemophilia B. Mol Biol Med 4: 11–20

Armentano D, Yu SF, Kantoff PW, von Ruden T, Anderson WF, Gilboa E (1987) Effect of internal viral sequences on the utility of retroviral vectors. J Virol 61: 1647–1650

Auffray C, Lillie JW, Korman AJ, Boss JM, Frechin N, Guillemot F, Cooper J, Mulligan RC, Strominger JL (1987) Structure and expression of HLA-DQ alpha and -DX alpha genes: interallelic alternate splicing of the HLA-DQ alpha gene and functional splicing of the HLA-DQ alpha gene using a retroviral vector. Immunogenetics 26: 63–73

Banapour B, Sernatinger J, Levy JA (1986) The AIDS-associated retrovirus is not sensitive to lysis or inactivation by human serum. Virology 152: 268–271

Bandyopadhyay PK, Temin HM (1984) Expression of a complete chicken thymidine kinase gene inserted in a retrovirus vector. Mol Cell Biol 4: 749–754

Bartholomew RM, Esser AF, Muller-Eberhard HJ (1978) Lysis of oncornaviruses by human serum. Isolation of the viral complement (CI) receptor and identification as p15E. J Exp Med 147: 844–853

Bender MA, Palmer TD, Gelinas RE, Miller AD (1987) Evidence that the packaging signal of Moloney murine leukemia virus extends into the gag region. J Virol 61: 1639–1646

Bender MA, Miller AD, Gelinas RE (1988) Expression of the human beta-globin gene after retroviral transfer into murine erythroleukemia cells and human BFU-E cells. Mol Cell Biol 8: 1725–1735

Bender MA, Gelinas RE, Miller AD (1989) A majority of mice show long-term expression of a human β-globin gene after retroviral transfer into hematopoietic stem cells. Mol Cell Biol 9: 1426–1434

Berger SA, Bernstein A (1985) Characterization of a retrovirus shuttle vector capable of either proviral integration or extrachromosomal replication in mouse cells. Mol Cell Biol 5: 305–312

Besmer P, Olshevsky D, Baltimore D, Dolberg D, Fan H (1979) Virus-like 30S RNA in mouse cells. J Virol 29: 1168–1176

Bosselman RA, Hsu R-Y, Bruszewski J, Hu F, Martin F, Nicolson M (1987) Replication-defective chimeric helper proviruses and factors affecting generation of competent virus: expression of Moloney murine leukemia virus structural genes via the metallothionein promoter. Mol Cell Biol 7: 1797–1806

Bowtell DD, Johnson GR, Kelso A, Cory S (1987) Expression of genes transferred to haemopoietic stem cells by recombinant retroviruses. Mol Biol Med 4: 229–250
Brown M, McCormack M, Zinn KG, Farrell MP, Bikel I, Livingston DM (1986) A recombinant murine retrovirus for simian virus 40 large T cDNA transforms mouse fibroblasts to anchorage-independent growth. J Virol 60: 290–293
Brown PO, Bowerman B, Varmus HE, Bishop JM (1987) Correct integration of retroviral DNA in vitro. Cell 49: 347–356
Cepko CL, Roberts BE, Mulligan RC (1984) Construction and applications of a highly transmissible murine retrovirus shuttle vector. Cell 37: 1053–1062
Chang SM, Wager-Smith K, Tsao TY, Henkel-Tigges J, Vaishnav S, Caskey CT (1987) Construction of a defective retrovirus containing the human hypoxanthine phosphoribosyltransferase cDNA and its expression in cultured cells and mouse bone marrow. Mol Cell Biol 7: 854–863
Collins SJ (1988) Different mechanisms account for the relative resistance of KG-1 and HL-60 cell lines to retrovirus infection. J Virol 62: 4346–4348
Cone RD, Mulligan RC (1984) High-efficiency gene transfer into mammalian cells: generation of helper-free recombinant retrovirus with broad mammalian host range. Proc Natl Acad Sci USA 81: 6349–6353
Cone RD, Reilly EB, Eisen HN, Mulligan RC (1987a) Tissue-specific expression of functionally rearranged lambda 1 Ig gene through a retrovirus vector. Science 236: 954–957
Cone RD, Weber-Benarous A, Baorto D, Mulligan RC (1987b) Regulated expression of a complete human beta-globin gene encoded by a transmissible retrovirus vector. Mol Cell Biol 7: 887–897
Cooper NR, Jensen FC, Welsh RM Jr, Oldstone MB (1976) Lysis of RNA tumor viruses by human serum: direct antibody-independent triggering of the classical complement pathway. J Exp Med 144: 970–984
Danos O, Mulligan RC (1988) Safe and efficient generation of recombinant retroviruses with amphotropic and ecotropic host ranges. Proc Natl Acad Sci USA 85: 6460–6464
Desiderio SV, Wolff KR (1988) Rearrangement of exogenous immunoglobulin VH and DJH gene segments after retroviral transduction into immature lymphoid cell lines. J Exp Med 167: 372–388
Dick JE, Magli MC, Huszar D, Phillips RA, Bernstein A (1985) Introduction of a selectable gene into primitive stem cells capable of long-term reconstitution of the hemopoietic system of W/W^v mice. Cell 42: 71–79
Dougherty JP, Temin HM (1987) A promoterless retroviral vector indicates that there are sequences in U3 required for 3′ RNA processing. Proc Natl Acad Sci USA 84: 1197–1201
Dougherty JP, Temin HM (1988) Determination of the rate of base-pair substitution and insertion mutations in retrovirus replication. J Virol 62: 2817–2822
Dzierzak EA, Papayannopoulou T, Mulligan RC (1988) Lineage-specific expression of a human beta-globin gene in murine bone marrow transplant recipients reconstituted with retrovirus-transduced stem cells. Nature 331: 35–41
Emerman M, Temin HM (1984a) Genes with promoters in retrovirus vectors can be independently suppressed by an epigenetic mechanism. Cell 39: 449–467
Emerman M, Temin HM (1984b) High-frequency deletion in recovered retrovirus vectors containing exogenous DNA with promoters. J Virol 50: 42–49
Emerman M, Temin HM (1986a) Quantitative analysis of gene suppression in integrated retrovirus vectors. Mol Cell Biol 6: 792–800
Emerman M, Temin HM (1986b) Comparison of promoter suppression in avian and murine retrovirus vectors. Nucleic Acid Res 14: 9381–9396
Feldman RA, Vass WC, Tambourin PE (1987) Human cellular fps/fes cDNA rescued via retroviral shuttle vector encodes myeloid cell NCP92 and has transforming potential. Oncogene Res 1:441–458
Franz T, Hilberg F, Seliger B, Stocking C, Ostertag W (1986) Retroviral mutants efficiently expressed in embryonal carcinoma cells. Proc Natl Acad Sci USA 83: 3292–3296
Garver RI Jr, Chytil A, Courtney M, Crystal RG (1987) Clonal gene therapy: transplanted mouse fibroblast clones express human alpha 1-antitrypsin gene in vivo. Science 237: 762–764
Gelinas C, Temin HM (1986) Nondefective spleen necrosis virus-derived vectors define the upper size limit for packaging reticuloendotheliosis viruses. Proc Natl Acad Sci USA 83: 9211–9215
Guild BC, Mulligan RC, Gros P, Housman DE (1988) Retroviral transfer of a murine cDNA for multidrug resistance confers pleiotropic drug resistance to cell without prior drug selection. Proc Natl Acad Sci USA 85: 1595–1599
Hartley JW, Rowe WP (1976) Naturally occurring murine leukemia viruses in wild mice: characterization of a new "amphotropic" class. J Virol 19: 19–25

Hawley RG, Covarrubias L, Hawley T, Mintz B (1987) Handicapped retroviral vectors efficiently transduce foreign genes into hematopoietic stem cells. Proc Natl Acad Sci USA 84: 2406–2410
Herman SA, Coffin JM (1987) Efficient packaging of readthrough RNA in ALV: implications for oncogene transduction. Science 236: 845–848
Hilberg F, Stocking C, Ostertag W, Grez M (1987) Functional analysis of a retroviral host-range mutant: altered long terminal repeat sequences allow expression in embryonal carcinoma cells. Proc Natl Acad Sci USA 84: 5232–5236
Hock RA, Miller AD (1986) Retrovirus-mediated transfer and expression of drug resistance genes in human haematopoietic progenitor cells. Nature 320: 275–277
Hogge DE, Humphries RK (1987) Gene transfer to primary normal and malignant human hemopoietic progenitors using recombinant retroviruses. Blood 69: 611–617
Holland CA, Anklesaria P, Sakakeeny MA, Greenberger JS (1987) Enhancer sequences of a retroviral vector determine expression of a gene in multipotent hematopoietic progenitors and committed erythroid cells. Proc Natl Acad Sci USA 84: 8662–8666
Hoshino H, Tanaka H, Miwa M, Okada H (1984) Human T-cell leukaemia virus is not lysed by human serum. Nature 310: 324–325
Hughes SH, Greenhouse JJ, Petropoulos CJ, Sutrave P (1987) Adaptor plasmids simplify the insertion of foreign DNA into helper-independent retroviral vectors. J Virol 61: 3004–3012
Huszar D, Balling R, Kothary R, Magli MC, Hozumi N, Rossant J, Bernstein A (1985) Insertion of a bacterial gene into the mouse germ line using an infectious retrovirus vector. Proc Natl Acad Sci USA 82: 8587–8591
Jahner D, Haase K, Mulligan R, Jaenisch R (1985) Insertion of the bacterial gpt gene into the germ line of mice by retroviral infection. Proc Natl Acad Sci USA 82: 6927–6931
Jat PS, Cepko CL, Mulligan RC, Sharp PA (1986) Recombinant retroviruses encoding simian virus 40 large T antigen and polyomavirus large middle T antigens. Mol Cell Biol 6: 1204–1217
Jhappan C, Vande Woude GF, Robins TS (1986) Transduction of host cellular sequences by a retroviral shuttle vector. J Virol 60: 750–753
Jolly DJ, Esty AC, Subramani S, Friedmann T, Verma IM (1983) Elements in the long terminal repeat of murine retroviruses enhance stable transformation by thymidine kinase gene. Nucleic Acids Res 11: 1855–1872
Kantoff PW, Gillio AP, McLachlin JR, Bordignon C, Eglitis MA, Kernan NA, Moen RC, Kohn DB, Yu SF, Karson E, Karlsson S, Zwiebel JA, Gilboa E, Blaese RM, Nienhuis A, O'Reilly RJ, Anderson WF (1987) Expression of human adenosine deaminase in nonhuman primates after retrovirus-mediated gene transfer. J Exp Med 166: 219–234
Kaplan PL, Simon S, Cartwright CA, Eckhart W (1987) cDNA cloning with a retrovirus expression vector: generation of a pp60c-src cDNA clone. J Virol 61: 1731–1734
Karlsson S, Papayannopoulou T, Schweiger SG, Stamatoyannopoulos G, Nienhuis AW (1987) Retroviral-mediated transfer of genomic globin genes leads to regulated production of RNA and protein. Proc Natl Acad Sci USA 84: 2411–2415
Karlsson S, Bodine DM, Perry L, Papayannopoulou T, Nienhuis AW (1988) Expression of the human beta-globin gene following retroviral-mediated transfer into multipotential hematopoietic progenitors of mice. Proc Natl Acad Sci USA 85: 6062–6066
Keller G, Paige C, Gilboa E, Wagner EF (1985) Expression of a foreign gene in myeloid and lymphoid cells derived from multipotent haematopoietic precursors. Nature 318: 149–154
King W, Patel MD, Lobel LI, Goff SP, Nguyen-Huu MC (1985) Insertion mutagenesis of embryonal carcinoma cells by retroviruses. Science 228: 554–558
Korman AJ, Frantz JD, Strominger JL, Mulligan RC (1987) Expression of human class II major histocompatibility complex antigens using retrovirus vectors. Proc Natl Acad Sci USA 84: 2150–2154
Kriegler M, Perez CF, Hardy C, Botchan M (1984) Transformation mediated by the SV40 T antigens: separation of the overlapping SV40 early genes with a retroviral vector. Cell 38: 483–491
Kuehn MR, Bradley A, Robertson EJ, Evans MJ (1987) A potential animal model for Lesch-Nyhan syndrome through introduction of HPRT mutations into mice. Nature 326: 295–298
Kwok WW, Schuening F, Steady RB, Miller AD (1986) Retroviral transfer of genes into canine hemopoietic progenitor cells in culture: a model for human gene therapy. Proc Natl Acad Sci USA 83: 4552–4555
Landau NR, Schatz DG, Rosa M, Baltimore D (1987) Increased frequency of N-region insertion in a murine pre-B-cell line infected with a terminal deoxynucleotidyl transferase retroviral expression vector. Mol Cell Biol 7: 3237–3243

Ledley FD, Darlington GJ, Hahn T, Woo SL (11987) Retroviral gene transfer into primary hepatocytes: implications for genetic therapy of liver-specific functions. Proc Natl Acad Sci USA 84: 5335–5339
Lemischka IR, Raulet DH, Mulligan RC (1986) Development potential and dynamic behavior of hematopoietic stem cells. Cell 45: 917–927
Lewis S, Gifford A, Baltimore D (1984) Joining of V kappa to J kappa gene segments in a retroviral vector introduced into lymphoid cells. Nature 308: 425–428
Linial ML, Miller AD (1990) Retroviral RNA packaging: sequence requirements and implications. Curr Top Microbiol Immunol 157: 125–152
Linney E, Neil SD, Prestridge DS (1987) Retroviral vector gene expression in F9 embryonal carcinoma cell. J Virol 61: 3248–3253
Lobel LI, Patel M, King W, Nguyen-Huu MC, Goff SP (1985) Construction and recovery of viable retroviral genomes carrying a bacterial suppressor transfer RNA gene. Science 228: 329–332
Lugo TG, Handelin B, Killary AM, Housman DE, Fournier RE (1987) Isolation of microcell hybrid clones containing retroviral vector insertions into specific human chromosomes. Mol Cell Biol 7: 2814–2820
Mann R, Baltimore D (1985) Varying the position of a retrovirus packaging sequence results in the encapsidation of both unspliced and spliced RNAs. J Virol 54: 401–407
Mann R, Mulligan RC, Baltimore D (1983) Construction of a retrovirus packaging mutant and its use to produce helper-free defective retrovirus. Cell 33: 153–159
Markowitz D, Goff S, Bank A (1988) A safe packaging line for gene transfer: separating viral genes on two different plasmids. J Virol 62: 1120–1124
Miller AD, Buttimore C (1986) Redesign of retrovirus packaging cell lines to avoid recombination leading to helper virus production. Mol Cell Biol 6: 2895–2902
Miller AD Jolly DJ, Friedmann T, Verma IM (1983) A transmissible retrovirus expressing human hypoxanthine phosphoribosyltransferase (HPRT): gene transfer into cells obtained from humans deficient in HPRT. Proc Natl Acad Sci USA 80: 4709–4713
Miller AD, Ong ES, Rosenfeld MG, Verma IM, Evans RM (1984) Infectious and selectable retrovirus containing an inducible rat growth hormone minigene. Science 225: 993–998
Miller AD, Law MF, Verma IM (1985) Generation of helper-free amphotropic retroviruses that transduce a dominant-acting, methotrexate-resistant dihydrofolate reductase gene. Mol Cell Biol 5: 431–437
Miller AD, Trauber DR, Buttimore C (1986) Factors involved in production of helper virus-free retrovirus vectors. Somat Cell Mol Genet 12: 175–183
Miller AD, Bender MA, Harris EAS, Kaleko M, Gelinas RE (1988) Design of retroviral vectors for transfer and expression of the human β-globin gene. J Virol 62: 4337–4345
Morgan JR, Barrandon Y, Green H, Mulligan RC (1987) Expression of an exogenous growth hormone gene by transplantable human epidermal cells. Science 237: 1476–1479
Morgan WC, Kaplan DR, Pallas DC, Roberts TM (1988) Recombinant retroviruses that transduce middle T antigen cDNAs derived from polyomavirus mutants: separation of focus formation and soft-agar growth in transformation assays and correlations with kinase activities in vitro. J Virol 62: 3407–3414
Murphy AJ, Efstratiadis A (1987) Cloning vectors vectors for expression of cDNA libraries in mammalian cells. Proc Natl Acad Sci USA 84: 8277–8281
Nelson DL, Weis JH, Przyborski MJ, Mulligan RC, Seidman JG, Housman DE (1984) Metaphase chromosome transfer of introduced selectable markers. J Mol Appl Genet 2: 563–577
Noda T, Satake M, Robins T, Ito Y (1986) Isolation and characterization of NIH 3T3 cells expressing polyomavirus small T antigen. J Virol 60: 105–113
Nolan GP, Fiering S, Nicolas JF, Herzenberg LA (1988) Fluorescence-activated cell analysis and sorting of viable mammalian cells based on beta-D-galactosidase activity after transduction of *Escherichia coli* lacZ. Proc Natl Acad Sci USA 85: 2603–2607
Osborne WRA, Miller AD (1988) Design of vectors for efficient expression of human purine nucleoside phosphorylase in skin fibroblasts from enzyme-deficient humans. Proc Natl Acad Sci USA 85: 6851–6855
Overell RW, Weisser KE, Cosman D (1988) Stably transmitted triple-promoter retroviral vectors and their use in transformation of primary mammalian cells. Mol Cell Biol 8: 1803–1808
Palmer TD, Hock RA, Osborne WR, Miller AD (1987) Efficient retrovirus-mediated transfer and expression of a human adenosine deaminase gene in diploid skin fibroblasts from an adenosine deaminase-deficient human. Proc Natl Acad Sci USA 84: 1055–1059

Palmer TD, Thompson AR, Miller AD (1989) Production of human factor IX in animals by genetically modified skin fibroblasts: potential therapy for hemophilia B. Blood 73: 438–445
Pastan I, Gottesman MM, Ueda K, Lovelace E, Rutherford AV, Willingham MC (1988) A retrovirus carrying an MDRI cDNA confers multidrug resistance and polarized expression of P-glycoprotein in MDCK cells. Proc Natl Acad Sci USA 85: 4486–4490
Price J, Turner D, Cepko C (1987) Lineage analysis in the vertebrate nervous system by retrovirus-mediated gene transfer. Proc Natl Acad Sci USA 84: 156–160
Rasheed S, Gardner MB, Chan E (1976) Amphotropic host range of naturally occurring wild mouse leukemia viruses. J Virol 19: 13–18
Reik W, Weiher H, Jaenisch R (1985) Replication-competent Moloney murine leukemia virus carrying a bacterial suppressor tRNA gene: selective cloning of proviral and flanking host sequences. Proc Natl Acad Sci USA 82: 1141–1145
Robertson E, Bradley A, Kuehn M, Evans M (1986) Germ-line transmission of genes introduced into cultured pluripotential cells by retroviral vector. Nature 323: 445–448
Rodland KD, Brown AMC, Magun B (1987) Individual mouse VL30 elements transferred to rat cells by viral pseudotypes retain their responsiveness to activators of protein kinase C. Mol Cell Biol 7: 2296–2298
Salter DW, Smith EJ, Hughes SH, Wright SE, Crittenden LB (1987) Transgenic chickens: insertion of retroviral genes into the chicken germ line. Virology 157: 236–240
Scolnick EM, Vass WC, Howk RS, Duesberg PH (1979) Defective retrovirus-like 30S RNA species of rat and mouse cells are infectious if packaged by type C helper virus. J Virol 29: 964–972
Sherwin SA, Rapp UR, Benveniste RE, Sen A, Todaro GJ (1978) Rescue of endogenous 30S retroviral sequences from mouse cells by baboon type C virus. J Virol 26: 257–264
Shih CC, Stoye JP, Coffin JM (1988) Highly preferred targets for retrovirus integration. Cell 53: 531–537
Shimotohno K, Temin HM (1981) Formation of infectious progeny virus after insertion of herpes simplex thymidine kinase gene into DNA of an avian retrovirus. Cell 26: 67–77
Shimotohno K, Temin HM (1982) Loss of intervening sequences in genomic mouse alpha-globin DNA inserted in an infectious retrovirus vector. Nature 299: 265–268
Sorge J, Hughes SH (1982) Splicing of intervening sequences introduced into an infectious retroviral vector. J Mol Appl Genet 1: 547–559
Sorge J, Wright D, Erdman VD, Cutting AE (1984) Amphotropic retrovirus vector system for human cell gene transfer. Mol Cell Biol 4: 1730–1737
Soriano P, Jaenisch R (1986) Retroviruses as probes for mammalian development: allocation of cells to the somatic and germ cell lineages. Cell 46: 19–29
Soriano P, Cone RD, Mulligan RC, Jaenisch R (1986) Tissue-specific and ectopic expression of genes introduced into transgenic mice by retroviruses. Science 234: 1409–1413
St Louis D, Verma IM (1988) An alternative approach to somatic cell gene therapy. Proc Natl Acad Sci USA 85: 3150–3154
Stacey A, Mulligan R, Jaenisch R (1987) Rescue of type I collagen-deficient phenotype by retroviral-vector-mediated transfer of human pro alpha I(I) collagen gene into Mov-13 cells. J Virol 61: 2549–2554
Stead RB, Kwok WW, Storb R, Miller AD (1988) Canine model for gene therapy: inefficient gene expression in dogs reconstituted with autologous marrow infected with retroviral vectors. Blood 71: 742–747
Stewart CL, Schuetze S, Vanek M, Wagner EF (1987) Expression of retroviral vectors in transgenic mice obtained by embryo infection. EMBO J 6: 383–388
Stoeckert CJ Jr, Metherall JE, Yamakawa M, Eisenstadt JM, Weissman SM, Forget BG (1987) Expression of the affected A gamma globin gene associated with Greek nondeletion hereditary persistence of fetal hemoglobin. Mol Cell Biol 7: 2999–3003
Stoker AW, Bissell MJ (1988) Development of avian sarcoma and leukosis virus-based vector-packaging cell lines. J Virol 62: 1008–1015
Stuhlmann H, Haenisch R, Mulligan RC (1989) Construction and properties of replication-competent murine retroviral vectors encoding methotrexate resistance. Mol Cell Biol 9: 100–108
Taketo M, Tanaka M (1987) A cellular enhancer of retrovirus gene expression in embryonal carcinoma cells. Proc Natl Acad Sci USA 84: 3748–3752
To RY, Booth SC, Neiman PE (1986) Inhibition of retroviral replication by anti-sense RNA. Mol Cell Biol 6: 4758–4762

Trevor K, Linney E, Oshima RG (1987) Suppression of endo B cytokeratin by its antisense RNA inhibits the normal coexpression of endo A cytokeratin. Proc Natl Acad Sci USA 84: 1040–1044
Turner DL, Cepko CL (1987) A common progenitor for neurons and glia persists in rat retina late in development. Nature 328: 131–136
Van Beveren C, Verma IM (1986) Homology among oncogenes. Curr Top Microbiol Immunol 123: 73–98
van der Putten H, Botteri FM, Miller AD, Rosenfeld MG, Fan H, Evans RM, Verma IM (1985) Efficient insertion of genes into the mouse germ line via retroviral vectors. Proc Natl Acad Sci USA 82: 6148–6152
Wagner EF, Vanek M, Vennstrom B (1985) Transfer of genes into embryonal carcinoma cells by retrovirus infection: efficient expression from an internal promoter. EMBO J 4: 663–666
Watanabe S, Temin HM (1983) Construction of a helper cell line for avian reticuloendotheliosis virus cloning vectors. Mol Cell Biol 3: 2241–2249
Wei CM, Gibson M, Spear PG, Scolnick EM (1981) Construction and isolation of a transmissible retrovirus containing the src gene of Harvey murine sarcoma virus and the thymidine kinase gene of herpes simplex virus type 1. J Virol 39: 935–944
Weis JH, Nelson DL, Przyborski MJ, Chaplin DD, Mulligan RC, Housman DE, Seidman JG (1984) Eukaryotic chromosome transfer: linkage of the murine major histocompatibility complex to an inserted dominant selectable marker. Proc Natl Acad Sci USA 81: 4879–4883
Weiss R, Teich N, Varmus H, Coffin J (1984) RNA tumor viruses. Cold Spring Harbor Laboratory, Cold Spring Harbor New York
Welsh RM Jr, Cooper NR, Jensen FC, Oldstone MB (1975) Human serum lyses RNA tumour viruses. Nature 257: 612–614
Welsh RM Jr, Jensen FC, Cooper NR, Oldstone MB (1976) Inactivation of lysis of oncornaviruses by human serum. Virology 74: 432–440
Wilson JM, Jefferson DM, Chowdhury JR, Novikoff PM, Johnston DE, Mulligan RC (1988a) Retrovirus-mediated transduction of adult hepatocytes. Proc Natl Acad Sci USA 85: 3014–3018
Wilson JM, Johnston DE, Jefferson DM, Mulligan RC (1988b) Correction of the genetic defect in hepatocytes from the Watanabe heritable hyperlipidemic rabbit. Proc Natl Acad Sci USA 85: 4421–4425
Wolff JA, Yee JK, Skelly HF, Moores JC, Respess JG, Friedmann T, Leffert H (1987) Expression of retrovirally transduced genes in primary cultures of adult rat hepatocytes. Proc Natl Acad Sci USA 84: 3344–3348
Yang Z, Korman AJ, Cooper J, Pious D, Accolla RS, Mulligan RC, Strominger JL (1987) Expression of HLA-DR antigen in human class II mutant B-cell lines by double infection with retrovirus vectors. Mol Cell Biol 7: 3923–3928
Yee JK, Moores JC, Jolly DJ, Wolff JA, Respess JG, Friedmann T (1987) Gene expression from transcriptionally disabled retroviral vectors. Proc Natl Acad Sci USA 84: 5197–5201
Yu SF, von Ruden T, Kantoff PW, Garber C, Seiberg M, Ruther U, Anderson WF, Wagner EF, Gilboa E (1986) Self-inactivating retroviral vectors designed for transfer of whole genes into mammalian cells. Proc Natl Acad Sci USA 83: 3194–3198

Poxvirus Expression Vectors

B. MOSS

1 Introduction

The poxviruses form a large family of complex DNA viruses that infect both vertebrate and invertebrate hosts. The most notorious member, variola virus, was responsible for smallpox, one of the most dreaded of all infectious diseases. In the late 18th century, Edward Jenner demonstrated that smallpox could be prevented by immunization with a more benign relative, vaccinia virus, which caused mild infections in cattle. Ultimately, global vaccination strategies led to the eradication of smallpox.

Expression of foreign genes by recombinant vaccinia was first described 7 years ago (MACKETT et al. 1982; PANICALI and PAOLETTI 1982). The possibility that such recombinant viruses containing genes from other microorganisms could be used for immunization against current diseases contributed greatly to the interest in vaccinia virus expression vectors (PAOLETTI et al. 1984; SMITH et al. 1983). Candidate human and veterinary live recombinant vaccines are

Laboratory of Viral Diseases, National Institute of Allergy and Infectious Diseases, National Institutes of Health, Bethesda, MD 20892, USA currently being tested. Thus far, however, the greatest use of recombinant vaccinia viruses has been made by immunologists and virologists to determine the targets of humoral and cell-mediated immune responses to microbial infections (Moss and Flexner 1987). The many advantages of poxvirus vectors, which include a wide host range, capacity for large amounts of foreign DNA, cytoplasmic site of transcription, and high expression levels, make them appropriate for many areas of research. In general, the recombinant-derived proteins appear to be properly processed and transported to the appropriate cellular compartment as dictated by their structure.

The popularity of the vaccinia virus expression system is at least partly due to the construction and wide dissemination of prototype plasmid insertion vectors that facilitated the construction and isolation of recombinant vaccinia viruses in laboratories with no poxvirus experience (Chakrabarti et al. 1985; Mackett et al. 1984). While the majority of vector work has been done with vaccinia virus, similar approaches have been used with certain other members of the poxvirus family, most notably fowlpox (Boyle and Coupar 1988a; Taylor et al. 1988a, b) and raccoonpox viruses (Esposito et al. 1988).

2 Molecular Biology of Poxviruses

A review of the molecular biology of poxviruses is available and contains the original references to most of the work described below (Moss 1985). Distinctive characteristics of members of the Poxvirus family include: (1) a large complex enveloped virion containing enzymes concerned with mRNA synthesis, (2) a genome composed of a single linear double-stranded DNA molecule of 130–300 kilobase pairs (kb), and (3) the ability to replicate within the cytoplasmic compartment of the cell. Infectious poxvirus particles contain a transcriptional system that can synthesize functional mRNA. To carry out this complex process, a large number of virus-encoded enzymes including RNA polymerase, capping and methylating enzymes, and poly(A) polymerase are packaged in the virus particle.

The genome of vaccinia virus is approximately 185 kb. Hairpin structures connecting the two strands are present at each end of the DNA. The sequence of about half of the genome has been reported and the remainder will probably be completed within the next year. Most of the essential genes occur within the central region of the DNA which is highly conserved among poxviruses; many of the genes that are dispensible for replication in tissue culture and those involved in host range are nearer the ends. The gene density is high and coding regions are typically separated by only a small number of bases. The apparent absence of introns, the relatively short promoter sequences, and the relatively small sizes of many open reading frames account for the packing of an estimated 150–200 genes into vaccinia virus.

The precise mechanism of virus entry into the cytoplasm is not well understood but it apparently involves fusion of the viral envelope with the plasma membrane. Within minutes after infection, viral mRNA synthesis that is programmed by the encapsidated enzymes can be detected. The genes encoding early mRNAs number about 100. Early promoters are about 30 base pairs (bp) long and can be divided into three regions: a 15-bp critical sequence, in which most single nucleotide substitutions have a major effect on expression, separated by 11 bp of a less critical usually T-rich sequence from a 5- to 7-bp region in which initiation usually occurs (DAVISON and MOSS 1991a). Like the TATA box of eukaryotic genes, the critical region specifies the site of transcription initiation. The sequence TTTTTNT, in which N can be any base including T, is found 20–50 bp upstream of the RNA termination site of early genes (YUEN and MOSS 1987). There is evidence that the signal is recognized in RNA as the corresponding UUUUUNU sequence and that the capping enzyme is involved in termination (SHUMAN et al. 1987; SHUMAN and MOSS 1988).

The synthesis of DNA polymerase and other early gene products permits the replication of poxvirus DNA to occur in the cytoplasm of infected cells. DNA replication signals the start of late gene expression, which can usually be detected at about 4 h after vaccinia virus infection, and the decline in expression of early genes. There appear to be about 100 late genes which encode the major structural polypeptides as well as some of the enzymes incorporated into virus particles. Late gene promoters differ in sequence from early promoters but may also be considered in terms of three regions: an upstream sequence of about 20 bp with some consecutive T or A residues, separated by a region of about 6 bp from a highly conserved TAAAT element within which transcription initiates (DAVISON and MOSS 1991b). Mutations within the TAAAT are detrimental and may eliminate expression entirely. Most commonly, a G residue follows the TAAAT generating the ATG initiation codon. Late mRNAs, however, contain a 5′ capped poly(A) leader of variable length that seems to be generated by slippage of the RNA polymerase near the start site (BERTHOLET et al. 1987; SCHWER et al. 1987). Late mRNAs are heterogeneous in length and the early transcription termination sequence is not recognized. Factors specific for late transcription have been partially purified (WRIGHT and MOSS 1989).

There are some vaccinia virus genes that have tandem early and late promoters providing for continuous expression (COCHRAN et al. 1985). In addition, there is evidence for a class of genes intermediate between early and late, but neither the critical elements of their promoter region nor the factors involved in their transcription have been determined (VOS and STUNNENBERG 1988).

With vaccinia virus, infectious progeny can usually be detected at about 6 h and continue for about 48 h yielding about 100–200 plaque forming units (2500 to 10000 particles). Assembly of virus particles is a complex process that can be inhibited by certain drugs such as rifampicin. With the commonly used strains of vaccinia virus, most of the infectious particles remain in the cytoplasm and must be released by cell lysis.

Infection of tissue culture cells with vaccinia virus and other orthopoxviruses results in profound cytopathic effects, changes in membrane permeability, and inhibition of host DNA, RNA, and protein synthesis. The effects on protein synthesis are dramatic and probably involve several factors; the relative contribution of each factor may depend on the virus multiplicity, cell type, and time of analysis. For some studies, it would be useful to have a mutant vaccinia virus that has good expression but minimal cytopathic effects. A variant with such properties, however, has not been demonstrated.

3 Production of Recombinant Viruses

3.1 General Strategy

The rational development and continued improvement of poxvirus vectors depends on an understanding of the molecular biology of this unique family. The large size of the poxvirus genome makes homologous recombination a practical way of inserting foreign genes. This is carried out by transfecting engineered plasmids into cells that have been infected with vaccinia virus (Fig. 1). In this manner, foreign DNA segments of up to 25000 bp have been incorporated into the vaccinia virus genome (Smith and Moss 1983). To retain

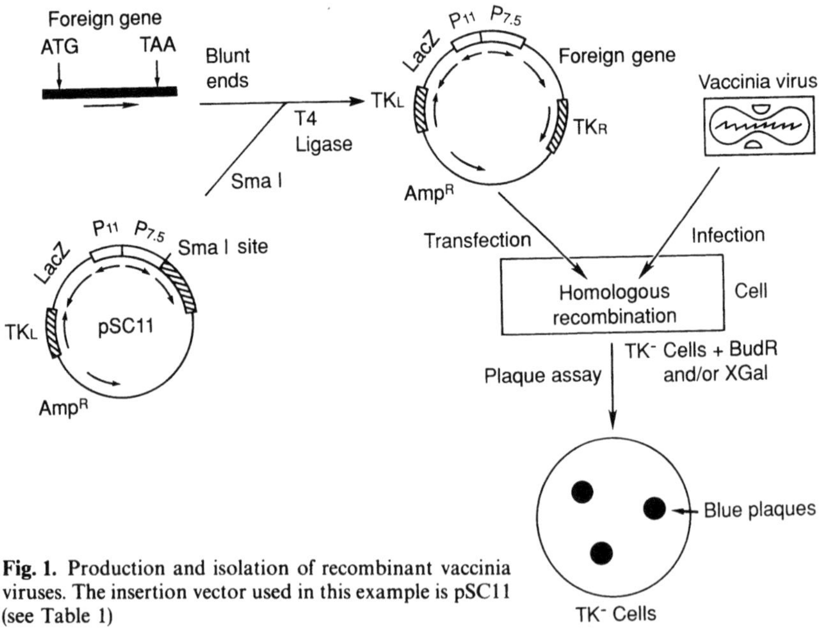

Fig. 1. Production and isolation of recombinant vaccinia viruses. The insertion vector used in this example is pSC11 (see Table 1)

Vector	Promoter[a]	Unique restriction sites	Flanking DNA	Selection/screening	Reference
pGS20	P7.5 (E/L)	*Bam*HI, *Sma*I[b]	TK	TK$^-$	MACKETT et al. 1984
pGS61	P7.5	*Bam*HI, *Hin*dIII	TK	TK$^-$	SMITH et al. 1987
pGS62	P7.5	*Bam*HI, *Sma*I, *Eco*RI	TK	TK$^-$	
pVV3	P7.5	Polylinker	TK	TK$^-$	RICE et al. 1985
pBCB01, 2, 3[c]	PF	Polylinker	TK	TK$^-$	BOYLE et al. 1989
pBCB06	P7.5	Polylinker	TK	TK$^-$	
pSC11	P7.5	*Sma*I[b]	TK	TK$^-$/β-gal	CHAKRABARTI et al. 1985
pSC11ss	P7.5	*Stu*I, *Sal*I	TK	TK$^-$/β-gal	EARL and MOSS (in preparation)
pCF11	P7.5	*Sma*I[b]	*Hin*dIII C	β-gal	FLEXNER et al. 1987
pYF6	P7.5	*Sma*I[b]	HA	HA$^-$, β-gal	
pPro 18	P7.5	*Sma*I[b]	HA	HA$^-$	SHIDA et al. 1987
pTK-7.5A	P7.5	Polylinker	*Hin*dIII F	TK$^+$	COUPAR et al. 1988
pTK-7.5B	P7.5	Polylinker	*Hin*dIII F	TK$^+$	
pUV I	P11 (L)	Polylinker[d]	TK	TK$^-$/β-gal	FALKNER et al. 1987
pTK-gpt-Fls, 2s, 3s	P11	Polylinker[e]	TK	TK$^-$ or gpt	FALKNER and MOSS 1988
pJ16[f]	P11 (E), P25 (L)	Multiple	TK	TK$^-$	TSAO et al. 1988
p1200	CAE I (L)	ClaI	TK	TK$^-$	PATEL et al. 1988
pMP528HRH	PH6 (E/L)	*Xho*I, *Kpn*I, *Sma*I[b]	*Hin*dIII K	Host range	PERKUS et al. 1989
pHES1, 2, 3	PH6	Polylinker[e]	*Hin*dIII K	Host range	
pHES4	PH6	Polylinker	*Hin*dIII K	Host range	
pTF7-5g	ϕ10 (T7)	*Bam*HI	TK	TK$^-$	FUERST et al. 1987
pT7·25EMC-1	ϕ10 + EMC	*Bam*HI[h]	TK	TK$^-$	ELROY-STEIN et al. 1989
pTM1[g]	ϕ10 + EMC	Polylinker[i]	TK	TK$^-$	MIZUKAMI, ELROY-STEIN, and MOSS, unpublished
pTM3[g]	ϕ10 + EMC	Polylinker[i]	TK	TK$^-$ or gpt	

[a] E, early; L, late; E/L, early and late; T7, phage T7; EMC, EMCV untranslated leader.
[b] *Sma*I digestion gives a blunt site for cloning any fragment after blunting.
[c] Initiation codons upstream of polylinker.
[d] Initiation codon immediately precedes *Eco*RI site
[e] Three-vector set with initiation codon followed by polylinker in all three open reading frames.
[f] Bidirectional promoters. P11 has consecutive initiation and termination codons.
[g] Used with vTF7-3, vaccinia virus that expresses T7 RNA polymerase (FUERST et al. 1986).
[h] EMCV initiation codon upstream of *Bam*HI site.
[i] Initiation codon is part of *Nco*I site

infectivity, however, the insertion must occur into silent regions of the genome or into nonessential genes. Since poxviruses have their own transcription system, the use of poxvirus promoters is obligatory. In addition, the foreign gene must consist of a continuous open reading frame since RNA splicing does not occur in the cytoplasm. If an early promoter is used to regulate the foreign gene, it is prudent to search for potential vaccinia virus transcription termination signls (TTTTTNT). The signal can be altered by in vitro mutagenesis without changing the coding sense and this may result in increased expression (P. Earl and B. Moss, unpublished observations).

To facilitate the construction of recombinant viruses, plasmid insertion vectors were developed (Mackett et al. 1984). These vectors contain a segment of vaccinia virus DNA within which an expression cassette, consisting of a vaccinia virus promoter followed by one or more unique restriction endonuclease sites for gene insertion, has been placed. The vaccinia virus DNA flanking the cassette serves to guide recombination to a nonessential site in the genome. The prototype vector, pGS20, and many derivatives and variants are listed in Table 1. Some of the features of these vectors will be pointed out in subsequent sections. Protocols describing methods of vaccinia virus production, recombination, and plaquing have been published (Mackett et al. 1985).

3.2 Transcriptional Regulatory Signals

The time and level of gene expression is dependent on the promoter chosen. An early poxvirus promoter will allow expression to occur before DNA replication, whereas expression will be delayed until after DNA replication if a late promoter is used. Because of the progressive cytopathic effects of vaccinia virus, early expression may sometimes be desirable. However, the strongest promoters belong to the late class and will generally provide higher level of protein. The most extensively used promoter, P7.5, is a compound promoter with both early and late transcription start sites, thereby providing for continued expression throughout the growth cycle (Cochran et al. 1985). The late P11 promoter (Bertholet et al. 1985) usually provides higher levels of expression than P7.5. A cowpox promoter, called CAE-1, or its vaccinia equivalent gives similar or slightly higher levels of expression than P11 (Patel et al. 1988). Still higher levels of expression have been achieved with a synthetic late promoter (Davison and Moss 1991b). The amounts of protein made are in the range of 1–2 μg per million cells.

Unique restriction endonuclease sites have been placed downstream of the promoter to facilitate gene insertion. The *Sma*I site gives a blunt end and therefore can be used as a universal site to clone any DNA fragment that has been made blunt by enzymátic treatment. Multiple restriction sites present in a polylinker have been added to some vectors for added convenience. Even with these vectors, however, it seems preferable to ligate the 5′ end of the gene as close to the promoter as possible in view of a report that translation is reduced by secondary structure produced by multiple copies of restriction sites (PELLETIER and SONENBERG 1985). The importance of placing a TTTTTNT transcription termination signal at the end of the foreign gene has not been critically evaluated. Nevertheless, a natural one is present at the end of the thymidine kinase (TK) gene which is present in the correct orientation relative to the foreign gene in many of the vectors employing early or early/late promoters listed in Table 1.

3.3 Insertion Site in the Vaccinia Virus Genome

As previously mentioned, the foreign gene must be inserted into a nonessential site if the recombinant virus is to retain infectivity. Many such sites have been identified, for example by sequence analysis of spontaneous deletion mutants (KOTWAL and MOSS 1988) or by making deletions or insertions (PERKUS et al. 1986). The most popular site has been the TK gene (HRUBY et al. 1983; WEIR and MOSS 1983) since, as will be discussed below, this provides for a method of selecting recombinant viruses. Insertion into the vaccinia hemagglutinin (HA) gene also provides a method of screening (SHIDA 1986). There have been no correlations of particular sites with stability or expression level. For laboratory work, the well characterized WR strain of vaccinia virus has been used most often. This virus grows to high titer, forms large plaques, exhibits a wide host range in vitro and in vivo, and is mouse adapted, making it convenient for experimentation. For vaccine development, either the New York City Board of Health (Wyeth) or Lister strains or more attenuated derivatives are recommended.

3.4 Selection and Screening of Recombinant Virus Plaques

Vaccinia virus makes large plaques on monolayers formed by a variety of cell lines; monkey kidney cells such as BSC-1 are particularly good. The plaques can be transferred to nitrocellulose and screened by hybridization to specific DNA sequences (PAOLETTI et al. 1984) or by binding to antibody (MACKETT and ARRAND 1985). Alternatively, a number of general selection and screening procedures have been adapted to poxvirus vectors.

The first and still most popular selection method is negative TK selection (MACKETT et al. 1982). The plasmid insertion vector contains the vaccinia virus TK gene within which the expression casette has been placed. When recombination occurs, the foreign DNA will recombine into the TK locus and interrupt the TK gene. Recombinant vaccinia viruses can then be distinguished from parental TK^+ virus by their TK^- phenotype. The basis for the selection is the lethal effect of incorporation of nucleoside analogs, such as 5-bromodeoxyuridine (BUdR), into the viral genome. Incorporation of the analog depends on its intracellular phosphorylation. TK^- cell lines (e.g., TK^- 143) must be employed so that phosphorylation is dependent on the TK of vaccinia virus. Therefore, TK^- virus will form BUdR-resistant plaques whereas TK^+ virus will not.

Although the selection between TK^- and TK^+ viruses is virtually complete, not every TK^- virus is a recombinant. Spontaneous TK^- mutant virus plaques appear at a frequency of 1:10,000. Depending on the transfection and recombination efficiency, the percentage of recombinants usually varies from 10% to 80% of the total TK^- virus but can be less if very large DNA segments are inserted. Therefore, it is necessary to screen the TK^- plaques. This can be done by infecting mini-well monolayers of cells with virus from plaques and then performing dot blot hybridization with extracted total DNA. Even more conveniently, DNA from a single plaque can be amplified directly by the polymerase chain reaction (PCR) and analyzed by gel electrophoresis (Y. Zhang, personal communication).

TK^+ selection is also possible by starting with a TK^- mutant of vaccinia virus. The herpes simplex virus TK gene is convenient to use since it has little sequence similarity with the vaccinia virus TK (Coupar et al. 1988; Mackett et al. 1982; Panicali and Paoletti 1982).

Another approach involves the cotransfer of a dominant selectable marker along with the desired foreign gene. This is accomplished by using vaccinia virus DNA to flank both the marker gene and the desired foreign gene. The prokaryotic neomycin-resistance (Franke et al. 1985) and guanine phosphoribosyltransferase (*gpt*) (Boyle and Coupar 1988b; Falkner and Moss 1988) genes, which provide resistance to G418 and mycophenolic acid, respectively, have been employed for this purpose. These selection systems permit the use of a wide variety of cell lines. With *gpt* selection, it has been possible to pick plaques directly without prior enrichment and without false positives as occurs with TK^- selection. For some purposes, it may not be desirable to actually incorporate the selectable marker into the stable recombinant vaccinia virus. To avoid this, a modification of the *gpt* selection procedure termed "transient dominant selection" (Falkner and Moss, in preparation) has been successfully employed. In this variation, the *gpt* gene is in the same plasmid as the desired foreign gene but is not flanked by vaccinia virus DNA. Nevertheless, the *gpt* gene (as well as the remainder of the plasmid) will be transiently incorporated into the vaccinia virus genome as a result of a single crossover event and mycophenolic acid-resistant plaques can be isolated. Since the single crossover results in a duplication of vaccinia virus sequences, however, the recombinant virus is unstable. Upon replaquing without selection, a mixture of wild-type and recombinant plaques are obtained. Thus, it is necessary to use either DNA hybridization or PCR to identify the recombinants. Since the *gpt* gene was not incorporated, it is possible to reuse this technique to insert additional genes into the same recombinant virus.

The finding that the deletion and replacement of a vaccinia virus gene modulates its host range in cultured human cells (Gillard et al. 1986) forms the basis for a host-range plaque selection system (Perkus et al. 1989). This system has an advantage in avoiding the use of any foreign genes for selection.

Several screening methods have been developed for picking recombinant virus plaques. One convenient method, illustrated in Fig. 1, involves the cotransfer of the β-galactosidase gene with the desired foreign gene (CHAKRABARTI et al. 1985; PANICALI et al. 1986). Recombinant plaques turn blue upon staining with a chromogenic β-galactosidase substrate. By using TK sequences for flanking the foreign DNA, it has been possible to combine TK selection and color screening. In this manner, it is easy to distinguish TK^- recombinants from spontaneous TK^- mutants. The reverse procedure, in which the β-galactosidase gene is first inserted into vaccinia virus and then recombinants are scored by "white" plaque formation has also been used. This procedure is not so convenient, however, because of false-positive white plaques.

Insertion into the vaccinia virus HA gene also provides a screening method since HA^+ plaques appear red upon addition of chicken erythrocytes whereas HA^- plaques do not (SHIDA 1986). Another screening method makes use of the observation that deletion of a gene encoding a 14000-kDa protein leads to the generation of small plaques (DALLO et al. 1987). By incorporating the wild-type version of this gene into the insertion vector, recombinants can be distinguished by their large plaque size (RODRIGUEZ and ESTEBAN 1989).

4 Other Poxvirus Vectors

The principles developed for the construction of vaccinia virus vectors can be employed for other members of the poxvirus family. Indeed, it is even possible to use vaccinia virus promoters in other poxvirus genera. The orthopoxviruses, including vaccinia virus, have a rather broad host range. By contrast, the members of some other poxvirus genera are more restricted in range. Thus far, fowlpox (BOYLE and COUPAR 1988a; TAYLOR et al. 1988a,b) and raccoon poxvirus (ESPOSITO et al. 1988) vectors have been tested.

5 Hybrid Vaccinia Virus/ Bacteriophage T7 Expression Systems

The bacteriophage T7 RNA polymerase is a single subunit enzyme, with high catalytic acitivity and strict promoter specificity, that has been widely used for in vitro transcription and for *Escherichia coli* expression vectors. By attaching a vaccinia virus promoter to the T7 RNA polymerase gene, recombinant vaccinia viruses that express T7 RNA polymerase have been constructed (FUERST et al. 1986). If a plasmid containing a gene regulated by a T7 promoter is transfected into a cell infected with the above recombinant vaccinia virus, relatively high level expression occurs. The level was estimated to be 400–600 times greater than that achieved with conventional nonreplicating plasmid mammalian transient expression vectors containing the enchancer and promoter regions of the Rous sarcoma virus long terminal repeat or the simian virus 40 early region.

Even greater expression and larger scale production was achieved by inserting the T7 promoter regulated gene into a second vaccinia virus and coinfecting cells with the latter and the virus containing the T7 RNA polymerase gene (Fuerst et al. 1987).

Although large amounts of T7 transcripts were made, with a steady state level equal to approximately 30% of the total cytoplasmic RNA, inefficient 5′ capping limited protein synthesis (Fuerst and Moss 1989). The cap requirement for translation was circumvented, however, by employing the untranslated RNA leader of encephalomyocarditis virus (EMCV) (Elroy-Stein et al. 1989). With this modification, the yield of chloramphenicol acetyltransferase (CAT) was estimated to be about 16.5 μg per 2 million cells or 10% of the total cell protein in 24 h. Pulse-labeling with [^{35}S]methionine revealed that CAT accounted for about 80% of the protein being synthesized at the latter time. The amount of protein made appears to be similar or higher than that achieved with the strong late poxvirus promoters.

The T7 system is a binary expression system, i.e., one virus carries the T7 RNA polymerase gene and another carries the T7 promoter regulated gene. It has not been possible to obtain both functions in a single viable virus (Fuerst et al. 1987). However, since expression does not occur while growing the individual virus stocks, the system should be capable of synthesizing toxic proteins that might otherwise interfere with virus yield.

6 Regulated Gene Expression

This *cis*- and *trans*-acting elements of the *E. coli lac* operon have been transferred to vaccinia virus and used to regulate gene expression (Fuerst et al. 1989). The *lac* repressor gene (*lacI*) was placed under control of the compound P7.5 promoter of vaccinia virus to ensure early and continuous production. It was calculated that each infected cell contained about 2×10^7 active repressor molecules. A strong vaccinia late promoter was modified by insertion of the *lac* operator (*lacO*) just downstream of the conserved TAAAT motif. Using the β-galactosidase gene as a test, it was found that little or no detectable enzyme was made in the absence of inducer; upon addition of β-D-thiogalactoside the yield was >20% of the unrepressed level. This inducible system might be especially valuable for expression of toxic proteins.

7 Uses of Recombinant Vaccinia Viruses

A high success rate has been obtained in expressing foreign genes using vaccinia virus vectors. Several factors undoubtedly contribute to this. Not least in importance are the cytoplasmic site of expression and the use of vaccinia transcription factors. The cytoplasmic site avoids potential problems related to cryptic splice sites, processing, and nuclear/cytoplasmic transport. For example, no problems were encountered in expressing the structural proteins of the human immunodeficiency HIV-1 without regard to the usual requirement for regulatory factors such as *rev* and *tat* (CHAKRABARTI et al. 1986; FLEXNER et al. 1988).

The proteins made by recombinant vaccinia viruses appear to be processed and transported in a manner similar if not identical to that occurring in uninfected cells. For example, N- and O-glycosylation (ELANGO et al. 1986), phosphorylation (GUY et al. 1987), proteolytic cleavage (CHAKRABARTI et al. 1986; RICE et al. 1985; THOSMAS et al. 1986), polarized plasma membrane (STEPHENS et al. 1986), and nuclear transport (SMITH et al. 1987; STOMATOS et al. 1987) all occur. Biologically active enzymes such as reverse transcriptase (FLEXNER et al. 1988) and bacteriophage T7 RNA polymerase (FUERST et al. 1986) are made.

The hybrid vaccinia/T7 system, or the improved EMCV version, is very powerful for transient expression. Since plasmids can be used for transfection, there is no need to make new recombinant vaccinia viruses. This approach is especially well suited for analytical studies, such as determining the effects of multiple mutations on the properties of a protein (MIZUKAMI et al. 1988). When scale-up is desired, however, the plasmids can be used as insertion vectors to make recombinant vaccinia viruses. The T7 system, prior to development of the EMCV modification, was already adapted to large scale purification of the envelope glycoprotein of HIV-1 using microcarrier cultures of Vero cell (BARRETT et al. 1989). The proteins produced by recombinant vaccinia viruses are active in the cells in which they are expressed. Thus, cells infected with a recombinant vaccinia virus that contains the HIV-1 envelope gene are able to fuse to cells bearing the CD4 receptor (LIFSON et al. 1986). Complementation experiments have also been successful; expression of the M (LI et al. 1988) and G (WHITT et al. 1989) proteins of vesicular stomatitis virus using the hybrid vaccinia/T7 system permitted the rescue of conditionally lethal temperature sensitive mutants of vesicular stomatitis virus. The success of the latter experiments depended on the compatability of the replication cycles of vaccinia virus and vesicular stomatitis virus.

Portability is another advantage of viral vectors, especially when the host range is extensive. Vaccinia virus can infect a large variety of mammals and birds as well as most cell lines derived from them. This property has been used to great advantage for immunological studies as alluded to at the start of this chapter and reviewed elsewhere (MOSS and FLEXNER 1987). Vaccinia virus vectors are now routinely employed for determining the targets of cytotoxic T cells and are being tested as candidate vaccines for human and veterinary purposes.

8 Precautions

Since poxviruses are infectious and can persist at ambient temperature when dried, good microbiological techniques must be followed. Ingestion, parenteral inoculation, and droplet or aerosol exposure of mucous membranes and broken skin are hazards to laboratory workers. Infection with vaccinia virus is rare, however, and in a generally healthy individual usually results in a localized vesicle at the site of penetration of the skin with a sharp object. Guidelines [HHS publication no. (CDC) 84-8395] written jointly by the Centers for Disease Control and the National Institutes of Health recommend Biosafety Level 2 practices, Class I or II biological safety cabinets, and vaccination at 3-year intervals. Vaccine has been available from the Centers for Disease Control.

References

Barrett N, Mitterer A, Mundt W, Eibl J, Eibl M, Gallo RC, Moss B, Dorner F (1989) Large-scale production and purification of a vaccinia recombinant-derived HIV-1 gp 160 and analysis of its immunogenicity. AIDS Res Hum Retroviruses 5: 159–171

Bertholet C, Drillien R, Wittek R (1985) One hundred base pairs of 5′ flanking sequence of a vaccinia virus late gene are sufficient to temporally regulate the transcription. Proc Natl Acad Sci USA 82: 2096–2100

Bertholet C, Van Meir E, ten Heggeler-Bordier B, Wittek R (1987) Vaccinia virus produces late mRNAs by discontinuous synthesis. Cell 50: 153–162

Boyle DB, Coupar BEH (1988a) Construction of recombinant fowlpox viruses as vectors for poultry vaccines. Virus Res 10: 343–356

Boyle DB, Coupar BEH (1988b) A dominant selectable marker for the construction of recombinant poxviruses. Gene 65: 123–128

Boyle DB, Coupar BEH, Both GW (1989) Multiple-cloning-site plasmids for the rapid construction of recombinant poxviruses. Gene 35: 169–177

Chakrabarti S, Brechling K, Moss B (1985) Vaccinia virus expression vector: Coexpression of β-galactosidase provides visual screening of recombinant virus plaques. Mol Cell Biol 5: 3403–3409

Chakrabarti S, Robert-Guroff M, Wong-Staal F, Gallo RC, Moss B (1986) Expression of the HTLV-III envelope gene by a recombinant vaccinia virus. Nature 320: 535–537

Cochran MA, Puckett C, Moss B (1985) In vitro mutagenesis of the promoter region for a vaccinia virus gene: evidence for tandem early and late regulatory signals. J Virol 54: 30–37

Coupar BEH, Andrew ME, Boyle DB (1988) A general method for the construction of recombinant vaccinia viruses expressing multiple foreign genes. Gene 68: 1–10

Dallo S, Rodriguez JF, Esteban M (1987) A 14K envelope protein of vaccinia virus with an important role in virus-host cell interactions is altered during virus persistence and determines the plaque size phenotype of the virus. Virology 159: 423–432

Davison AJ, Moss B (1991a) The structure of vaccinia virus early promoters. J Mol Biol (in press)

Davison AJ, Moss B (1991b) The structure of vaccinia virus late promoters. J Mol Biol (in press)

Elango N, Prince GA, Murphy BR, Venkatesan S, Chanock RM, Moss B (1986) Resistance to human respiratory syncytial virus (RSV) infection induced by immunization of cotton rats with a recombinant vaccinia virus expressing the RSV G glycoprotein. Proc Natl Acad Sci USA 83: 1906–1911

Elroy-Stein O, Fuerst TR, Moss B (1989) Cap-independent translation of mRNA conferred by encephalomyocarditis virus 5′ sequence improves the performance of the vaccinia virus/bacteriophage T7 hybrid expression system. Proc Natl Acad Sci USA 86: 6126–6130

Esposito JJ, Knight JC, Shaddock JH, Novembre FJ, Bauer GM (1988) Successful oral rabies vaccination of raccoons with raccoon poxvirus recombinants expressing rabies virus glycoprotein. Virology 167: 313–316

Falkner FG, Moss B (1988) *Escherichia coli gpt* gene provides dominant selection for vaccinia virus open reading frame expression vectors. J Virol 62: 1849–1854

Falkner GG, Chakrabarti S, Moss B (1987) pUV I: a new vaccinia virus insertion and expression vector. Nucleic Acids Res 15: 7192

Flexner C, Hugin A, Moss B (1987) Prevention of vaccinia virus infection in immunodeficient nude mice by vector-directed IL-2 expression. Nature 330: 259–262

Flexner C, Broyles SS, Earl P, Chakrabarti S, Moss B (1988) Characterization of human immunodeficiency virus gag/pol gene products expressed by recombinant vaccinia viruses. Virology 166: 339–349
Franke CA, Rice CM, Strauss JH, Hruby DE (1985) Neomycin resistance as a dominant selectable marker for selection and isolation of vaccinia virus recombinants. Mol Cell Biol 5: 1918–1924
Fuerst TR, Moss B (1989) Structure and stability of mRNA synthesized by vaccinia virus-encoded bacteriophage T7 RNA polymerase in mammalian cells. Importance of the 5′ untranslated leader. J Mol Biol 206: 333–348
Fuerst TR, Niles EG, Studier FW, Moss B (1986) Eukaryotic transient expression system based on recombinant vaccinia virus that synthesizes bacteriophage T7 RNA polymerase. Proc Natl Acad Sci USA 83: 8122–8126
Fuerst TR, Earl PL, Moss B (1987) Use of a hybrid vaccinia virus T7 RNA polymerase system for expression of target genes. Mol Cell Biol 7: 2538–2544
Fuerst TR, Fernandez MP, Moss B (1989) Transfer of the inducible *lac* repressor/operator system from *Escherichia coli* to a vaccinia virus expression vector. Proc Natl Acad Sci USA 86: 2549–2553
Gillard S, Spehner D, Drillien R, Kirn A (1986) Localization and sequence of a vaccinia virus gene required for multiplication in human cells. Proc Natl Acad Sci USA 83: 5573–5577
Guy B, Kieny MP, Riviere Y, Le Peuch C, Dott K, Girard M, Montagnier L, Lecocq J-P (1987) HIV F/3′ orf encodes a phosphorylated GTP-binding protein resembling an oncogene product. Nature 330: 266–269
Hurby DE, Maki RA, Miller DB, Ball LA (1983) Fine structure analysis and nucleotide sequence of the vaccinia virus thymidine kinase gene. Proc Natl Acad Sci USA 80: 3411–3415
Kotwal GJ, Moss B (1988) Analysis of a large cluster of nonessential genes deleted from a vaccinia virus terminal transposition mutant. Virology 167: 524–537
Li Y, Luo L, Snyder RM, Wagner RR (1988) Expression of the M gene of vesicular stomatitis virus cloned in various vaccinia virus vectors. J Virol 62: 776–782
Lifson JD, Feinberg MB, Reyes GR, Rabins L, Bonapour B, Chakrabarti S, Moss B, Wong-Staal F, Steimer KS, Engelman EG (1986) Induction of CD4-dependent cell fusion by the HTLV-III/LAV envelope glycoprotein. Nature 323: 725–728
Mackett M, Arrand JR (1985) Recombinant vaccinia virus induces neutralizing antibodies in rabbits against Epstein-Barr virus membrane antigen gP340. EMBO J 4: 3229–3234
Mackett M, Smith GL, Moss B (1982) Vaccinia virus: a selectable eukaryotic cloning and expression vector. Proc Natl Acad Sci USA 79: 7415–7419
Mackett M, Smith GL, Moss B (1984) General method for production and selection of infectious vaccinia virus recombinants expressing foreign genes. J Virol 49: 857–864
Mackett M, Smith GL, Moss B (1985) The construction and characterization of vaccinia virus recombinants expressing foreign genes. DNA cloning. IRL Press, Oxford
Mizukami T, Fuerst TR, Berger EA, Moss B (1988) Binding region for human immunodeficiency virus (HIV) and epitopes for HIV-blocking monoclonal antibodies of the CD4 molecule defined by site-directed mutagenesis. Proc Natl Acad Sci USA 85: 9273–9277
Moss B (1985) Replication of poxviruses. In: Virology. Raven, New York
Moss B, Flexner C (1987) Vaccinia virus expression vectors. Ann Rev Immunol 5: 305–324
Panicali D, Paoletti E (1982) Construction of poxviruses as cloning vectors: insertion of the thymidine kinase gene from herpes simplex virus into the DNA of infectious vaccinia virus. Proc Natl Acad Sci USA 79: 4927–4931
Panicali D, Grzelecki A, Huang C (1986) Vaccinia virus vectors utilizing the β-galactosidase assay for rapid selection of recombinant viruses and measurement of gene expression. Gene 47: 193–199
Paoletti E, Lipinskas BR, Samsonoff C, Mercer SR, Panicali D (1984) Construction of live vaccines using genitically engineered poxviruses: biological acitivity of vaccinia virus recombinants expressing the hepatitis B virus surface antigen and the herpes simplex virus glycoprotein D. Proc Natl Acad Sci USA 81: 193–197
Patel DD, Ray CA, Drucker RP, Pickup DJ (1988) A poxvirus-derived vector that directs high levels of expression of cloned genes in mammalian cells. Proc Natl Acad Sci USA 85: 9431–9435
Pelletier J, Sonenberg N (1985) Insertion mutagenesis to increase secondary structure within the 5′ noncoding region of a eukaryotic mRNA reduces translation efficiency. Cell 40: 515–526
Perkus ME, Panicali D, Mercer S, Paoletti E (1986) Insertion and deletion mutants of vaccinia virus. Virology 152: 285–297
Perkus ME, Limbach K, Paoletti E (1989) Cloning and expression of foreign genes in vaccinia virus using a host range selection system. J Virol 63: 3829–3836

Rice CM, Franke CA, Strauss JH, Hruby DE (1985) Expression of Sindbis virus structural proteins via recombinant vaccinia virus: synthesis, processing, and incorporation into mature Sindbis virions. J Virol 56: 227–239
Rodriguez JF, Esteban M (1989) Plaque size recombinants as a selectable marker to generate vaccinia virus recombinants. J Virol 63: 997–1001
Schwer B, Visca P, Vos JC, Stunnenberg HG (1987) Discontinuous transcription or RNA processing of vaccinia virus late messengers results in a 5′ poly(A) leader. Cell 50: 163–169
Shida H (1986) Nucleotide sequence of the vaccinia virus hemagglutinin gene. Virology 150: 451–462
Shida H, Tochikura T, Sato T, Konno T, Hirayoshi K, Seki M, Ito Y, Hatanaka M, Hinuma Y, Sugimoto M, Takahashi-Nisimaki F, Maruyama T, Miki K, Suzuki K, Morita M, Sashiyama H, Hayami M (1987) Effect of the recombinant vaccina viruses that express HTLV-I envelope gene on HTLV-I infection. EMBO J 6: 3379–3384
Shuman S, Moss B (1988) Factor-dependent transcription termination by vaccinia virus RNA polymerase: evidence that the cis-acting termination signal is in nascent RNA. J Biol Chem 263: 6220–6225
Shuman S, Broyles SS, Moss B (1987) Purification and characterization of a transcription termination factor from vaccinia virions. J Biol Chem 262: 12372–12380
Smith GL, Moss B (1983) Infectious poxvirus vectors have capacity for at last 25,000 base pairs of foreign DNA. Gene 25: 21–28
Smith GL, Mackett M, Moss B (1983) Infectious vaccinia virus recombinants that express hepatitis B antigen. Nature 302: 490–495
Smith GL, Levin JZ, Palese P, Moss B (1987) Synthesis and cellular location of the ten influenza polypeptides individually expressed by recombinant vaccinia viruses. Virology 160: 336–345
Stephens EB, Compans RW, Earl P, Moss B (1986) Surface expression of viral glycoproteins is polarized in epithelial cells infected with recombinant vaccinia virus vectors. EMBO J 5: 237–245
Stomatos N, Chakrabarti S, Moss B, Hare DJ (1987) Expression of polyomavirus virion proteins by a vaccinia virus vector: association of VP1 and VP2 with the nuclear framework. J Virol 61: 516–525
Taylor J, Weinberg R, Kawaoka Y, Webster RG, Paoletti E (1988a) Protective immunity against avian influenza induced by a fowlpox virus recombinant. Vaccine 6: 504–508
Taylor J, Weinberg R, Languet B, Desmettre P, Paoletti E (1988b) Recombinant fowlpox virus inducing protective immunity in nonavian species. Vaccine. 6: 497–503
Thomas G, Herbert E, Hruby DE (1986) Expression and cell typespecific processing of human preproenkephalin with a vaccinia recombinant. Science. 232: 1641–1643
Tsao H, Liu G-Q, Ruan L, Chu C-M (1988) Construction and application of plasmids containing bidirectional promoters of vaccinia virus. J Virol 62: 4832–4834
Vos JC, Stunnenberg HG (1988) Derepression of a novel class of vaccinia virus genes upon DNA replication. EMBO J 7: 3487–3492
Weir JP, Moss B (1983) Nucleotide sequence of the vaccinia virus thymidine kinase gene and the nature of spontaneous frameshift muations. J Virol 46: 530–537
Whitt MA, Chong L, Rose JK (1989) Glycoprotein cytoplasmic domain sequences required for rescue of a vesicular stomatitis virus glycoprotein mutant. J Virol 63: 3569–3578
Wright CF, Moss B (1989) Identification of factors specific for transcription of the late class of vaccinia virus genes. J Virol 63: 4224–4233
Yuen L, Moss B (1987) Oligonucleotide sequence signaling transcriptional termination of vaccinia virus early genes. Proc Natl Acad Sci USA 84: 6717–6421

Expression of Heterologous Sequences in Adenoviral Vectors

K. L. Berkner

1 Introduction

Adenoviruses (Ad) have served as outstanding agents in contributing to an understanding of transformation and regulation of gene expression in cultured cells (reviewed in Tooze 1981; Doerfler 1983, 1984; Ginsberg 1984). Three decades after their original isolation (Rowe et al. 1953) 41 serotypes of Ad which cause infections in over a dozen species have been identified. Initial classification was based on immunological criteria and on turmorigenicity in immunocompetent rodents, which ranges from highly oncogenic (e.g., Ad12) to generally nononcogenic (e.g., Ad2 and Ad5). More recently DNA homology has also been included as a criterion. Because they provided the first example of a human virus capable of inducing tumors in experimental animals and because Ad is capable of transformation in vitro, there was obvious, immediate interest in them. Subsequent revelation that the Ad heavily employs much of the cellular machinery during its life cycle made the adenovirus family an excellent probe

ZymoGenetics, 4225 Roosevelt Way N. E., Seattle WA 98105, USA for dissecting macromolecular biosynthesis and expanded this interest. Adaptation of Ad as a vector for various applications was thus a natural consequence of the considerable wealth of information accrued about this virus. Establishing cell lines by transformation with Ad recombinants that contain heterologous genes was one obvious application. Another was its development as a vector for producing high level expression of heterologous genes, stimulated by the extremely efficient lytic expression of this virus. Ad has also been tested for potential suitability in vaccine development, as well as for use in analyzing tissue-specific events. Since the rationale in these experimental designs stems largely from the information compiled on the Ad life cycle, this will be reviewed first, and then specific applications will be discussed. Although Ad vectors also have been heavily exploited for studying Ad biology per se, this review will be limited to describing the adaptation of Ad for the expression of heterologous sequences.

2 Ad Biology

2.1 Lytic Cycle

The Ad lytic cycle is biphasic, comprising an early and a late phase demarcated by the onset of replication. Promoters from seven regions of the Ad genome are known to be operational at early times, and most of these transcribe early (E) regions 1–4 (Fig. 1; for reviews, see BERK 1986; LEVINE 1984). In general, each early region is composed of multiple transcripts containing common 5′ and 3′ termini which are distinguished by differential internal splicing. Over 30 early transcripts have been identified; however, functions ascribed to the translation products are limited to a small subset of these. In several cases the pattern of transcription within the early regions changes dramatically at late times.

The Ela gene products, the first viral proteins synthesized after infection, are the major Ad transcriptional regulators and are also involved in DNA synthesis induction and transformation (reviewed by NEVINS 1987; GRAND 1987). The 289 aa Ela protein is the primary protein that facilitates transcription of the remaining early promoters. The 289 aa and 243 aa Ela proteins are capable, as well, of transactivating cellular promoters and of transrepressing both viral and cellular promoters. The Ela region is sufficient for the immortalization of primary cells and can cooperate with the Elb region or with other oncogenes in the transformation of primary cultures. The creation of a human cell line (293) that contains and expresses most of the Ela and Elb regions (GRAHAM et al. 1977) has been invaluable both in the genetic analysis of these regions and in the complementation of Ad recombinants with heterologous sequences substituted for El.

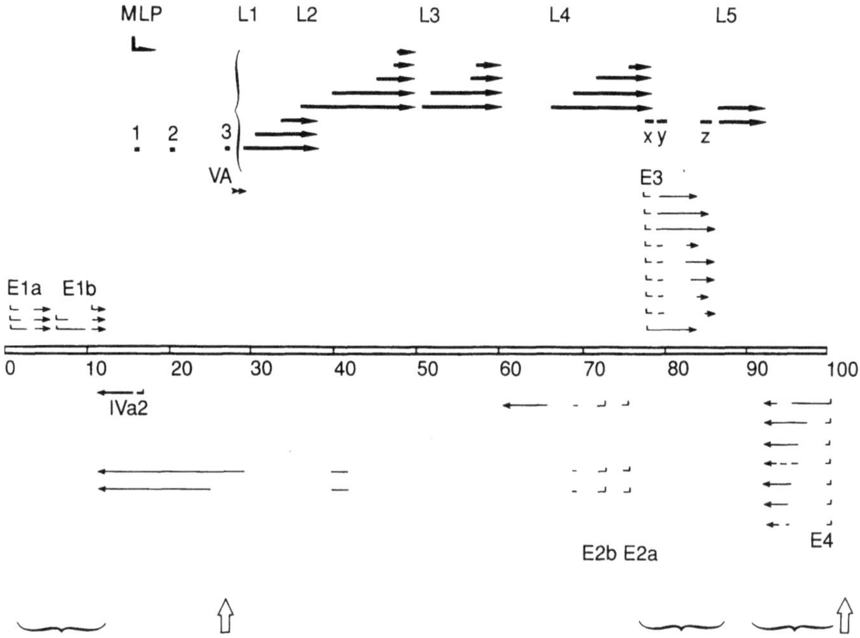

Fig. 1. Transcriptional map of an adenovirus (Ad). The linear Ad genome is represented by 0–100 map units (mu) where each unit equals 360 bp. Early transcripts are indicated by *light lines* and the five late families (*L1–L5*) by *heavy lines*. The 5′ mRNA termini are represented by *vertical notches* and the *arrowheads* indicate the 3′ termini. The E2 promoters active at early (E2a, 76 mu) and late (E2b, 72 mu) times are both indicated. The *vertical bracket* represents the alternative splicing of individual late messages from L1 to L5 onto the tripartite leader, which is shown here by *squares* labeled 1–3. The *brackets* and *vertical arrows* at the bottom of the figure denote possible regions for substitution or insertion, respectively, of heterologous sequences in helper-independent Ad recombinants

Besides functioning to effect full transformation, the 55 kDa and 19 kDa proteins in the Elb region are also responsible for cytoplasmic accumulation of late viral mRNA, for protection of viral and cellular DNA sequences from nucleolytic degradation, and for the shutoff of host message transport and translation (Babiss et al. 1985; Babiss and Ginsberg 1984; Pilder et al. 1984, 1986; Lai Fatt and Mak 1982; Subramanian et al. 1984; White et al. 1984). Additionally, there is a late Elb transcript that originates from a distinct promoter after the onset of DNA replication which encodes a virion component, pIX. pIX has been found to be required for packaging full-length Ad genomes and is, therefore, important for generating most substituted Ad recombinants (Ghosh-Choudhury et al. 1987). Elb deletions propagated in 293 cells, which do not express pIX, produce thermolabile virions (Colby and Shenk 1981).

The E4 region, like the Elb region, is important in effecting the early to late transition, specifically in late viral gene expression and in host protein synthesis shutoff, and E4 is also necessary for viral replication and virion assembly

(HALBERT et al. 1985; YODER and BERGET 1986; WEINBERG and KETNER 1986; FALGOUT and KETNER 1987; SANDLER and KETNER 1989; BRIDGE and KETNER 1989). This region is complex and probably the least understood, although the creation of an E4-encoding cell line (W162) should contribute to an accelerated understanding (WEINBERG and KETNER 1983). There may be as many as 24 E4 mRNAs, and combined sequence and cDNA analysis has so far identified seven open reading frames (ORFs). Deletion and insertion mutations disrupting six of these ORFs still generate viable virus, suggesting that the gene products overlap and are functionally compensatory (HALBERT et al. 1985). More recent experiments indicate that ORFs 3 and 6 are primarily responsible for the E4 phenotype (BRIDGE and KETNER 1989). It will be interesting, however, to see whether these mutations effect the same viral phenotype in cells from different tissues. In vitro translation suggests a potential for as many as 16 polypeptides derived from E4. Only a few E4 polypeptides have been identified in vivo however. One of these, a 34 kDa polypeptide encoded by ORF 6, is found associated in vivo with the 55 kDa protein encoded by the Elb region. Since both Elb and E4 mutations share the same complex phenotype, the 55 kDa-34 kDa complex may represent the functional unit affecting the early to late transition during lytic infection (SARNOW et al. 1984; CUTT et al. 1987).

The E3 region is believed to participate in immunosurveillance in vivo, as the two activities so far identified from this region have been shown to be involved in immune defense events. Sequence analysis has identified nine messages in E3 (Fig. 1) with the potential for translating as many proteins, and three of these (19 kDa, 50 kDa, 14.7 kDa) have been detected in vivo (PERSSON et al. 1980; WOLD et al. 1984; TOLLEFSON and WOLD 1988). The 19 kDa glycoprotein appears to protect Ad-infected cells against class I-restricted cytotoxic T cell lysis. This glycoprotein, which contains a C-terminal sequence targeting it to the endoplasmic reticulum, complexes intracellularly with HLA/H-2 class I histocompatibility antigens (PÄÄBO et al. 1983, 1987; SIGNÄS et al. 1982; KÄMPE et al. 1983; KVIST et al. 1978). This association inhibits both terminal glycosylation and cell-surface expression of the histocompatibility antigens, presumably reducing recognition of the infected cell by the cellular immune system (ANDERSSON et al. 1985; BURGERT and KVIST 1985; BURGERT et al. 1987). The other E3 protein for which an activity has been assigned, 14.7 kDa in size, is responsible for suppression of cytolysis induced by tumor necrosis factor (GOODING et al. 1988). Although important in vivo, the E3 region has been known for some time to be dispensable for growth in cultured cells, both from the isolation of viable Ad mutants containing substitutions for E3 and from the ability to delete most of the E3 sequences while still retaining viability (ANDERSON et al. 1976; BERKNER and SHARP 1983; GLUZMAN et al. 1982; JONES and SHENK 1978; KELLY and LEWIS 1973).

The E2 region originates from two promoters, one at 76 map units (mu) that operates early and one at 72 mu that is active only at late times. This region, important in both viral transformation and replication, encodes a 72 kDa DNA binding protein, an 80 kDa preterminal protein, and a 40 kDa DNA polymerase (reviewed by KELLY 1984). The function of the 72 kDa protein is poorly understood; besides being required for replication and transformation it may play a regulatory role, as it appears to specifically repress E4 transcription (NEVINS and WINKLER 1980). The 80 kDa preterminal protein is the precursor for the 55 kDa protein found covalently attached to each terminus of the Ad genome. Ad replication initiates with the formation of an ester bond between dCMP and a ser in the 80 kDa protein, which subsequently serves as a primer for DNA replication at the Ad termini. Each end consists of an identical, inverted, terminal repeat (ITR) of approximately 100 bp, depending on the serotype, which is required for replication. DNA synthesis proceeds from either end by single-strand displacement to generate both a duplex- and a single-stranded DNA product. The latter then undergoes another round of replication. At the later states of infection, probably during virion maturation, the 80 kDa protein is processed to the 55 kDa form, which remains attached to each terminus.

The availability of an accurate, efficient in vitro replication system has facilitated the analysis of the requirements for viral replication (reviewed by STILLMAN 1985). In addition to the E2 gene products described here, at least three host factors have been identified as being required for DNA replication. ITR sequences have been shown to be the only *cis* sequences required for replication. When ITR sequences cloned into plasmids are transfected into cells, they function as linear, autonomously replicating minichromosomes, dependent upon transacting factors provided by coinfection with helper virus (HAY et al. 1984; HAY 1985a, b; HAY and MCDOUGALL 1986). Both this approach and the in vitro replication system have been valuable in delineating minimal ITR sequence requirements for replication.

A shift from the early to late stage occurs approximately 8 h postinfection, marked by the onset of viral replication and concomitant with the inhibition of host message transport and translation. Translation shutoff is uncoupled from message transport inhibition, as mRNAs already in the cytoplasm, as well as messages that escape the transport block, still cannot be translated (BABICH et al. 1983; KHALILI and WEINMANN 1984; MOORE et al. 1987). Late stage events also include an alteration in early promoter activities, new transcription from two late promoters (producing pIX and IVa_2), and qualitative and quantitative changes in expression from the major late promoter (MLP). Efficient replication results in roughly a 1000 fold increase in genome copy number. Concomitantly, transcription from the MLP, which occurs at only low levels early in infection, undergoes an approximately 1000-fold increase. The transcription pattern generated by the MLP at late times is also altered: in contrast to early times when the L1 family accounts for most of the MLP-generated mRNA, transcription now proceeds from about 17 mu to 99 mu, with subsequent alternative splicing of each transcript giving rise to one of five different late mRNA families (L1–5, see Fig. 1). Members of each family are 3′ coterminal, and all of these main body messages are appended at their 5′ terminus to a tripartite leader (BERGET et al. 1977; CHOW et al. 1977). This sequence, named for the three leader elements spliced from Ad sequences at mu 17, 20 and 27, is thought to be responsible for effecting efficient translation of Ad messages at late times (LOGAN and SHENK 1984; BERKNER and SHARP 1985) and they have been adapted for efficient expression of heterologous genes, as will be described in Sect. 5.

Another sequence important for translation of Ad messages at late times is VAI RNA (THIMMAPPAYA et al. 1982). This small (approximately 160 nucleotide) RNA polymerase III product is synthesized at early times but the level of transcription increases dramatically during the late stage of infection. In addition to being required for translation of Ad messages at late times, VAI RNA also increases the translation of cellular messages (SVENSSON and AKÜSJARVI 1985). This suggests that it operates by a general mechanism, which appears to be its ability to restore the activity of a host initiation factor, eIF2 (O'MALLEY et al. 1986; AKÜSJARVI et al. 1987; KATZE et al. 1987 and references therein).

At least 30 mRNAs are generated by alternative splicing of the tripartite leader onto the various messages of the five late families. These transcripts encode structural proteins involved in virion assembly. Several of the proteins are synthesized in precursor form, with subsequent processing by a 19 kDa virion endopeptidase (CHATTERJEE and FLINT 1987 and references therein). After approximately 48 h the lytic cycle is complete, resulting in cell death and a burst size for Ad of approximately 10^3–10^4 pfu/cell, generating virus titers of 10^9–10^{10} pfu per ml.

The extensive knowledge gathered thus far about events during Ad infection has been largely limited to analysis in only a few human cell lines (mostly HeLa, VERO, KB, and 293 cells). Not surprisingly, differences in expression in other cell types have been reported, such as the effect of Elb deletions upon late expression in lymphocytes versus fibroblasts (RUETHER et al. 1986). Additional, analysis has been restricted predominantly to a small subset of serotypes (Ad2, Ad5, Ad12). Variance in oncogenicity alone signifies that Ad is not one functionally generic group. Differences in tissue-specific viral expression will be of obvious interest in ultimately sorting out viral functions in vivo, but are mentioned here as a caveat to bear in mind in appropriately adapting Ad as a viral vector.

2.2 Transformation

The most intensely analyzed Ad-transformed cells have been rodent-derived, although human and simian lines are also transformed by Ad (for reviews, see GRAHAM 1984b; GRAHAM, et al. 1984; WILLIAMS 1986). Ad transformation is characterized by stable integration of viral sequences into the cellular genome with no detectable free or rescuable virus. This is in contrast to lytic infection, where integration is not detected. All viral-generated transformants retain and express the left end 12%–14% of the viral DNA, and DNA fragments encoding even less than this fraction of the genome (i.e., the left 8%) are sufficient for effecting full transformation. Viral-mediated transformation is also influenced by non-E1 sequences. Adenoviral variants mutated in the E2 region, specifically in the genes encoding the 140 kDa DNA polymerase and the 72 kDa DNA binding protein, are altered in their ability to transform rodent cells. A single point mutation in the 140 kDa DNA polymerase gene was shown to be responsible for a defect in the initiation but not in the maintenance of transformation (MILLER and WILLIAMS 1987). The E2 region does not appear to be required for replication, since transformation can occur under conditions where replication is blocked. Possibly the E2 region facilitates transformation through complex formation of the Ad genome with E2 gene products (GRAHAM 1984b).

Most virus-generated transformants contain integrated subgenomic DNA fragments. However, cells containing all or nearly all of the genome, integrated such that the structure is colinear with the input viral DNA and sometimes repeated several times in a tandem arrangement, have also been isolated. These structures have been observed as well with recombinant virus-generated transformants, as will be described in Sect. 8. Retention of most of the genome correlates with conditions for nonpermissive viral growth, e.g., mutations in E1 or infection of a nonpermissive host. When these cell lines have been analyzed by subcloning and sequencing the integrated Ad DNA, deletions such as the loss of ITR sequences are observed, accounting for the failure to rescue infectious virus from these lines. Another structure frequently observed in transformed lines is left and right termini joined together. These observations have been the basis for proposing that circularization of the Ad genome, possibly facilitated by the E2 gene products, occurs prior to integration (GRAHAM et al. 1984; RUBEN et al. 1983). Covalently closed Ad circles have been observed in Ad-infected human and rodent cells before the onset of replication, providing support for this model (RUBEN et al. 1983). Subsequent rearrangement of integrated sequences could then generate the different general structures described.

DNA-mediated transformation most likely operates by an independent mechanism, since neither E2-encoded functions nor circle formation would be involved in a transfection with only the left end 8% of the genome. Additionally, viral-mediated transformation is much more efficient than DNA-mediated transformation. This feature has made using Ad vectors attractive for developing cell lines, as will be described in a subsequent section.

3 Advantages and Limitations in Using Ad Vectors

Information accumulated on Ad biology suggested several potential ways in which Ad might be valuable for vector development and actual implementation has confirmed its value. Extensive knowledge of Ad transcriptional organization has provided some basis for making the appropriate adaptations. For example, insertion of heterologous sequences within the E1 or E3 regions, under control of the respective endogenous early promoters, has been used to obtain expression of foreign genes. Alternatively, expression cassettes which incorporate the MLP have been effective for heterologous gene expression. The potent strength of the MLP, coupled with the ability of Ad to shut off cellular protein synthesis at late times, have made Ad vectors particularly valuable where overexpression of proteins for biochemical analysis is desired. Additionally, the availability of 293 cells adapted for growth in suspension culture facilitates scale-up capabilities (GRAHAM 1987). When heterologous genes are expressed in Ad recombinants, the appropriate post-translational processing and subcellular targeting are observed.

The complex Ad transcriptional program is still far from understood and occasionally transcription of heterologous sequences has been unpredictably affected by surrounding Ad sequences, a feature which will be detailed in subsequent sections. Another limitation is that wild-type (wt) Ad infection ultimately kills the host cell, thereby limiting the time scale for protein production. Vectors with appropriate Ad transcriptional and translational elements have been constructed that are capable of effecting high-level protein production in transfected cells without causing death of the host (KAUFMAN 1985; POWELL et al. 1986). For the production of secreted proteins, where host translation shut-off is not such an advantage, a cell line generated with these vectors is often preferable to infection with viral vectors. An exception to this would be the expression of a toxic gene product.

The Ad genome is large ($\sim$ 36 kb) and current helper-independent vectors have the capacity to accept up to approximately 7.5 kb of foreign DNA. This results from the ability to package genomes of up to 106 mu (i.e., 38 kb), from the availability of cell lines that express and thus complement the E1 or E4 regions, and from the dispensability of the E3 region in vitro. There is no known lower packaging limit, offering greater flexibility in the construction of recombinants. To date, the largest reported insert in an Ad recombinant is 6 kb in size (GHOSH-CHOUDHURY et al. 1987). Presumably, with helper-dependent recombinants even larger substitutions could be made. The only essential *cis* active sequences are the ITRs and a packaging signal found at the left end of the genome, which can function at the right end as well (HEARING and SHENK 1983; HEARING et al. 1987). Theoretically, a helper-dependent recombinants consisting of only the ITRs and packaging signal could accommodate inserts of up to 37 kb. This approach, however, has not been developed.

Adenoviruses exhibit a broad host range specificity. They replicate to low levels in simian and mouse cells and to high titers in a wide variety of human cells. Their efficient penetration (100% in many of the cell lines) enhances their suitability both in the analysis of heterologous gene expression as well as in optimizing protein production in infected cells. Since individual cells can absorb multiple virions, expression of more than one product is possible. The observed ability of exogenous promoters inserted within the Ad genome to effect transcription (GAYNOR et al. 1984; HEARING and SHENK 1985) has made Ad vectors amenable for studying tissue specificity, especially since analysis is possible in a large variety of cell types. Exogenous transcriptional termination signals embedded within the Ad genome have also been found to be functional (FALCK-PEDERSEN et al. 1985). Adenoviruses transform a variety of human, simian, and rodent cells. Their efficient transformation frequency makes them an attractive alternative to other methods of DNA transfer in some cell types and provides a useful option for creating a cell line that contains a desired gene product. The efficiency of Ad transformation is still well below that of retroviral vectors, probably limiting their use in gene therapy.

Finally, Ad recombinants can be made with relative ease, in part because of their high frequency of recombination in vivo (CHINNADURAI et al. 1979). Once generated, the Ad recombinants have been found to be quite stable, with very little observed rearrangement. This includes Ad recombinants that contain reiterated sequences, such as a duplicated MLP, within the same genome.

4 Strategies for Generating Ad Recombinants

Generation of helper-independent Ad recombinants can be accomplished by insertion of heterologous sequences into several different regions of the Ad genome. Most of these adaptations have been made in Ad5, although some vector development has also been done with Ad2, Ad4, and Ad7. Substitution of the E1 or E4 regions produces conditional recombinants which can be propagated in complementing cell lines (GRAHAM et al. 1977; WEINBERG and KETNER 1983). Nonconditional helper-independent Ad recombinants can be prepared by insertion into or substitution of the E3 region, by insertion of heterologous sequences either between the E4 promoter and the right ITR (SAITO et al. 1985), or possibly between the sequences encoding the DNA polymerase and 80 kDa terminal protein in the E2 region (MUNZ and YOUNG 1987). Foreign genes have either been inserted into the Ad genome as part of a self-sufficient expression cassette that includes a promoter, or they have been substituted or inserted within the E1 or E3 region downstream of the respective early promoters. Genes inserted in either transcriptional orientation within E1 or E3 are expressed, even though with E3 inserts one orientation opposes the direction of Ad late transcription (see Fig. 1). the first E1 deletions constructed extended to 11 mu. However, retention of sequences between 9 and 11 mu is desirable, given the requirement for pIX in packaging full-length viral genomes (GHOSH-CHOUDHURY et al. 1987). Substitution for E3 is conventionally performed by insertion at a unique XbaI site which is generated after deletion of the sequences between 79 and 84 mu (see Fig. 1). An expression cassette inserted here, in the left to right transcriptional orientation, positions the foreign gene behind both the MLP and E3 promoters, and at least one of the these appears to influence the transcription pattern. When an Ad recombinant containing the SV40 early promoter in an expression cassette was substituted for E3, the SV40

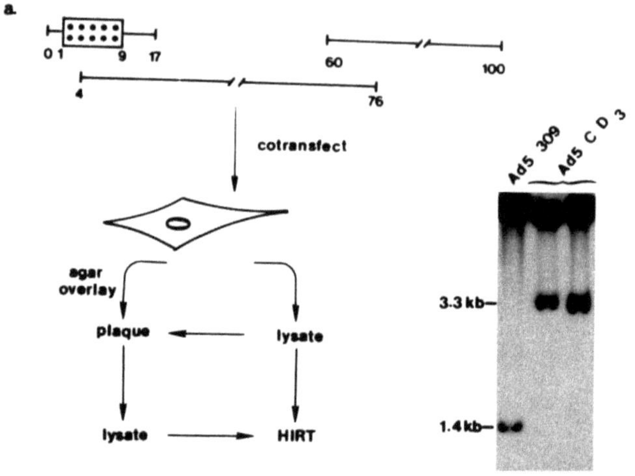

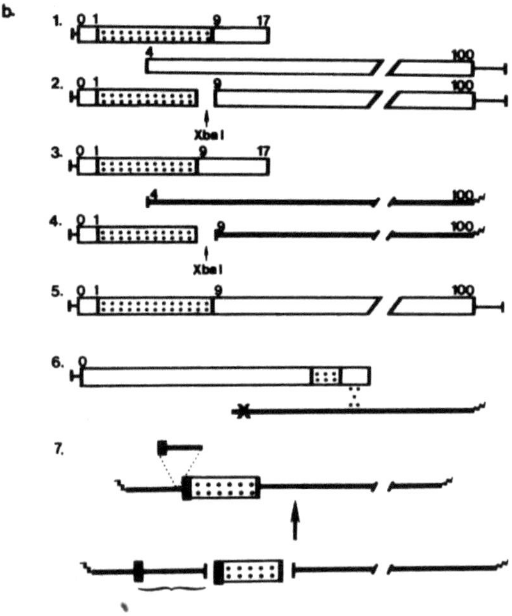

Fig. 2 a, b. Strategies for generating Ad recombinants. The procedure for producing Ad recombinants is generalized in **a**, along with a representative gel illustrating the screen used for progeny. In vivo ^{32}P-labeled DNA is Hirt extracted and cleaved, here with *Xba*I, to distinguish two independent isolates of recombinant Ad DNA [Ad5(C – D)$_3$] from the parental wild type (wt) Ad DNA [Ad5(309)] used in creating the recombinant. The appearance of a 3.3 kb *Xba*I fragment indicates the presence of the desired recombinant virus. **b** Indicates the various approaches used to produce recombinants. Ad plasmid sequences are represented by *open boxes* and foreign sequences by *dotted boxes*. The *thin lines* indicate bacterial vector sequences and the *heavy lines* represent viral DNA. The *squiggle* at the viral DNA termini indicates residual aa's from the 55 kDa terminal protein that remain after DNA isolation. Individual strategies depicted here are described in the text. Number *6* represents the use of conditionally defective viral DNA in selecting for appropriate recombination;

promoter was shown by SI analysis to be silent (JOHNSON et al. 1988). Transcription and translation of the foreign gene were observed however, presumably due to the MLP and/or E3 promoter activity.

There are a number of approaches for producing Ad recombinants. Plasmids that contain the entire genome are infectious, as are plasmids containing subgenomic fragments with overlapping sequence that together comprise the total genome (Fig. 2b1, 2, 5; BERKNER and SHARP 1983; HANAHAN and GLUZMAN 1984; GRAHAM 1984a). Reconstructing the full-lenth Ad genome with these subgenomic fragments is accomplished via in vitro ligation or by homologous recombination in vivo following transfection (Fig. 2b1, 2). The amount of overlap required between plasmids can be as small as 1 kb (BERKNER and SHARP 1983). Although sequences along the Ad genome may not all be euqlly recombinogenic, to date this has not appeared to be a limitation in constructing viral recombinants. Subgenomic plasmid sequences can also be joined to purified viral DNA fragments by ligation or recombination (Fig. 2b3, 4). This approach was used to construct the first Ad recombinant (STOW 1981). The main limitation with combining viral and plasmid fragments lies in the potential background from contaminating full-length viral DNA. Such background can increase the amount of work required in screening progeny and can be significant if the recombinant under construction exhibits reduced viability compared with the cognate viral DNA. Incorporating the use of conditionally defective viral or plasmid DNA fragments can help considerably to reduce background (GHOSH-CHOUDHURY et al. 1987; BERKNER 1988; MCGRORY et al. 1988).

The large size of the Ad genome unfortunately precludes there being many unique, useful restriction sites available for in vitro manipulation. Selection for Ad genomes containing reduced numbers of individual restriction sites has been useful in generating Ad variants more amenable to manipulation (JONES and SHENK 1978, 1979). In particular this approach has made construction of recombinants substituted at the left end quite straightforward. After in vitro manipulation of a left-end plasmid, the sequences are joined to viral DNA at a unique *Xba*I (4 or 9 mu) or *Cla*I (9 mu) site. However, there are still only a handful of convenient unique restriction sites for the rest of the genome, and so in general the easiest way to create Ad recombinants is by homologous recombination using subgenomic Ad plasmid DNA fragments (Fig. 2b1). This approach also has the advantage of avoiding background problems.

With full-length or subgenomic plasmids with the left and right termini of Ad joined to pBR322 sequences, cleavage between the bacterial vector and the left or right terminus is required for infectivity (BERKNER and SHARP 1983; HANAHAN and GLUZMAN 1984). In contrast, a full-length Ad plasmid with pML

◄

the *X* represents a lethal marker in the viral DNA. Represented in 7 is the insertion via ligation of a heterologous sequence (*lower drawing*) followed by intramolecular homologous recombination resulting in the deletion of Ad sequences (*upper drawing*). The *heavy vertical bars* represent homologous Ad sequences. The region deleted during homologous intramolecular recombination is indicated both by a *bracket* in the lower drawing and as the structure highlighted by the *dotted lines* in the upper drawing inserted within the El region was found to be infectious in circular form (GRAHAM 1984a). Linker sequences attached to the Ad termini are removed during viral growth, resulting in progeny with DNA termini that are indistinguishable from that of progenitor wild-type (wt) Ad DNA (BERKNER and SHARP 1983; HANAHAN and GLUZMAN 1984). The efficiency of totally cloned sequences in generating Ad recombinants is considerably lower (i.e., approximately 100-fold less) than that observed using plasmid joined to viral sequence, which in turn is about 25-fold less efficient than wt Ad. However, in most cases the efficiency is sufficient to generate the desired viral recombinant.

Helper-dependent Ad recombinants have been generated using a completely different approach (Fig. 2b7; THUMMEL et al. 1982, 1983). Foreign genes flanked at the 5′ end by Ad DNA are inserted into the genome via ligation at a convenient restriction site, and subsequent in vivo intramolecular recombination between the flanking sequence and homologous region in Ad DNA targets the inserted sequence to the desired position in the Ad genome. Because essential regions of the genome are deleted helper virus is required, as is some selection to maintain the desired recombinant in the population. To date this has been accomplished by inserting the SV40 T antigen (TAg) gene into Ad. Following propagation in human cells, the recombinant virus is enriched in CV1 cells which are not permissive for Ad unless the gene is present.

The process of acquiring Ad recombinants (Fig. 2a) involves transfecting an appropriate cell line (usually 293 cells) and isolating plaques directly from the transfected cells, or harvesting the transfected cells after 2 days and then plaque-purifying the virus. Analysis of the viral DNA by Hirt extraction (HIRT 1967) is also useful. The latter approach offers the advantage that one can quickly determine the approximate percentage of desired recombinants that are present in the virus stock, providing an indication of how many plaques it will be necessary to screen. Figure 2a illustrates the type of restriction analysis used to distinguish background virus from desired progeny.

5 Overproduction of Heterologous Proteins Using Ad Vectors

High level production of foreign proteins expressed in Ad vectors has been accomplished either by using MLP-based expression cassettes substituted for the E1 region or by incorporating a gene into the Ad genome downstream of the endogenous MLP. The former approach has the advantage of creating a helper-independent recombinant where 100% of the infecting Ad population is the desired recombinant. The latter approach generates a helper-dependent vector in which the percentage of recombinants can range between a few percent and 30 percent, a number which unfortunately is not predictable. Embedding a gene behind the natural MLP, however, may take greater advantage of all endogenous transcriptional—and translational—regulatory sequences, which might be missed in constructing an expression cassette. Both approaches have in some cases achieved high levels of heterologous gene expression, and have been reciprocally instructive in designing optimized expression vectors.

The ectopic MLP in the helper-independent Ad recombinant expression cassettes, which has included between 200 and 700 bp upstream of the transcriptional start site, is efficient, and temporally regulated during infection. Levels of heterologous mRNAs are comparable to those of individual, abundant Ad late mRNAs. The levels, however, should in theory equal the sum of all the late mRNAs, and the fact that this has not been observed indicates that the cassettes lack sequences required for full transcriptional activity. At least one of these sequences is in the intron between the first and second leader, as will be described. Expression of the heterologous gene is observed at late times. With several recombinants tested, expression has not been observed during the early phase (BERKNER and SHARP 1984, 1985; BERKNER et al. 1987; DAVIDSON and HASSELL 1987). This differs from the minimal early transcriptional activity observed for the endogenous MLP, suggesting that the 200 bp upstream of the MLP probably lack the sequences necessary for expression at early times. However, the temporal expression of heterologous genes has curiously been found to depend upon sequences downstream of the promoter. An identical expression cassette yielded transcription of thymidine kinase (tk) but not of polyoma middle T antigen (pymT) at early times (DAVIDSON and HASSELL 1987; MOORE and SHENK 1988).

Most expression cassettes have included at least 172 of the 202 nucleotide (nt) tripartite leader. As mentioned in Sect. 2, this leader is found appended to all of the late structural messages, which are translated so efficiently. Juxtaposing the tripartite leader in front of nonviral sequences increases their translation as well; 20-fold or 5-fold increases in translation efficiency were observed when dihydrofolate reductase (DHFR) or EIa, respectively, were positioned behind the first leader or a cDNA encoding the tripartite leader (LOGAN and SHENK 1984; BERKNER and SHARP 1985). In another construction in which the hepatitis B surface antigen gene (HBsAG) was inserted behind either the first leader or a tripartite leader-encoding cDNA, a 70-fold difference was observed. Since the constructions were not otherwise identical, however, an absolute comparison is not possible (DAVIS et al. 1985). The length of the sequence between the leader and initiation codon can have a significant effect upon translation. With DHFR expression more efficient translation was observed when this distance was short, a situation that mimics the actual structures so far determined for several of the late messages (BERKNER et al. 1987). This position requirement may account for the observed failure of the tripartite leader to enhance the translation of genes encoding tk or SV40 TAg where the distance between the leader and initiation codon was about 50 nt (MANSOUR et al. 1986; MOORE and SHENK 1988). Interestingly, in one of these studies the tripartite leader was shown to play a role in mRNA stability by promoting both a decreased nuclear- and increased cytoplasmic-mRNA half-life (MOORE and SHENK 1988).

The role of the tripartite leader in facilitating translation is not understood. The sequence predicts a highly structured leader, yet translation of tripartite leader-containing mRNAs is resistant to increases in ionic strength in both infected cells and in cell-free lysates (CHERNEY and WILHELM 1979). Although Ad messages are capped, this modification does not appear to be important for efficient translation. Late message translation is unaffected by coinfection with poliovirus, which results in the cleavage of p220, a component of the CAP-binding complex (CASTRILLO and CARRASCO 1987; DOLPH et al. 1988). Moreover, when the tripartite leader is embedded internally in mRNA, completely precluding CAP modification at its 5′ end, it still facilitates efficient translation (BERKNER et al. 1988). A matched set of Ad recombinants encoding polycistronic messages for chloramphenicol acetyl transferase (CAT) and murine DHFR were constructed with the tripartite leader positioned upstream of either the first cistron or both cistrons (Fig. 3b). Although efficient translation of the first cistron (CAT) was observed in both cases, efficient translation of the second cistron (DHFR) was observed only when a tripartite leader was positioned just upstream of it. Thus the tripartite leader facilitates translation only of the cistron immediately downstream of it, and can operate independently of the 5′ terminus and CAP modification. The anomalies in the effect of hypertonicity upon translation and in the apparent CAP independence of the tripartite leader have raised questions as to whether this sequence enhances translation by the conventionally accepted scanning model. The recent observation that picornaviral sequences can facilitate internal initiation (JANG et al. 1988; PELLETIER and SONENBERG 1988) clearly allows for alternative mechanisms.

A strong polyadenylation (pA) signal is important for optimal heterologous gene expression. Both the Elb and SV40 pA signals have worked well, while the polyoma (py) pA signal does not appear to be as strong. Inefficient pA was observed in two different Ad recombinants with a pymT cDNA containing the py pA, substituted in both orientations in El (DAVIDSON and HASSELL 1987). Heterologous mRNA populations were observed with each recombinant . This included, in one orientation, pA at the downstream Elb site. Cryptic splice site utilization was also reported. This is in contrast to what was obtained with a different pymT cDNA-encoding Ad recombinant which contained an SV40 pA, and splice signals; efficient termination was observed and the only pymT mRNA synthesized was the one predicted from the transcriptional signals incorporated into the expression cassette (BERKNER et al. 1987).

Another aberrant processing event that has been observed in an Ad recombinant involved the splice signals included in an expression cassette (BERKNER and SHARP 1984). In an El-substituted cassette expressing DHFR, similar to that shown in Fig. 3b but containing the first leader and its 5′ splice site (ss), efficient splicing was observed to the downstream Elb 3′ ss. However, in an Ad recombinant that differed only in containing the entire tripartite leader and the third leader 5′ ss, splicing into Elb was not observed. Avoiding aberrant transcriptional processing events is clearly important for preventing the dilution of functional transcripts. On the other hand, incorporating splice signals into an expression cassette may be required for expression. In one experiment, substantially less mRNA was observed in a recombinant with an expression cassette

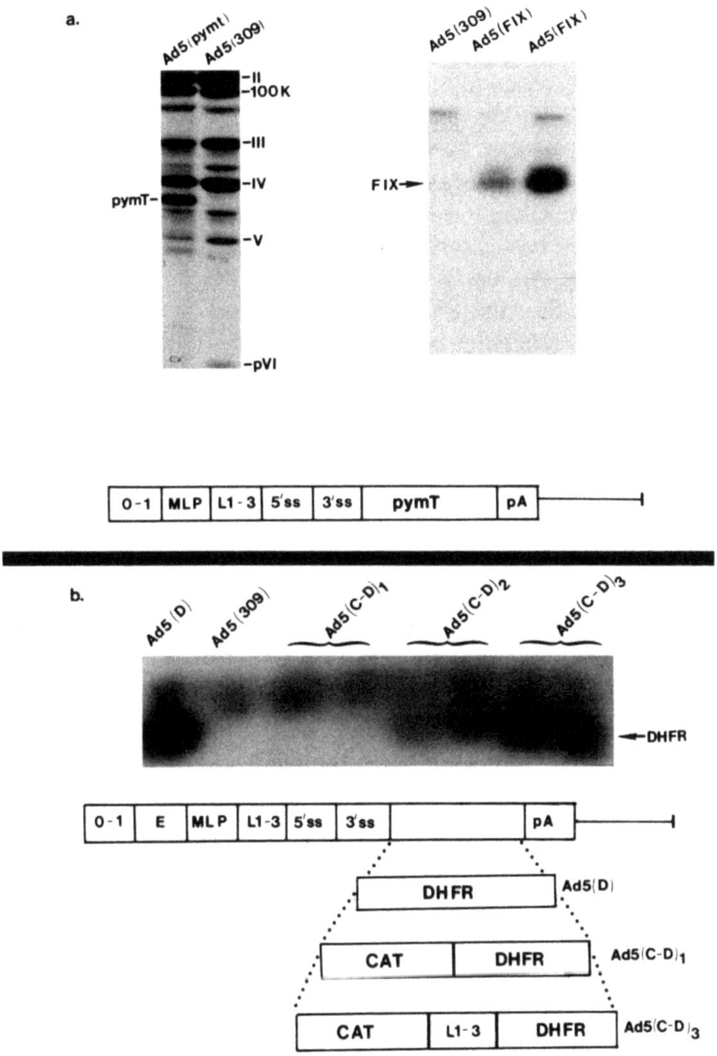

Fig. 3 a, b. Expression of heterologous proteins with Ad vectors. **a** The expression of an intracellular (pymT) and secreted (factor IX or FIX) protein is compared with wt virus, Ad5 (309). Infected cells were labeled with ^{35}S-met (*pymT*) or ^{35}S-cys (FIX) for 1 h at late times (20–24 h post infection). The cell pellet (pymT) or cell supernatant (FIX) was then harvested and analyzed using gel electrophoresis. The vector used in expressing one of these proteins (pymT) is shown below this analysis. **b** Cells infected with wild-type (wt) virus or the indicated recombinants were harvested 20 h post infection and DHFR levels were determined by western analysis using an antibody against murine DHFR and ^{125}I-labeled protein A. The upper mol. wt. band, present even in wt-infected cells, is probably human DHFR. Ad5(C − D)$_2$ is a polycistronic encoding recombinant that contains the tripartite leader only upstream of DHFR. The vector in **b** contains the SV40 enhancer sequences (*E*) upstream of the Ad MLP lacking splice site (BERKNER and SHARP 1985). To avoid alternative splicing in the case of El-substituted recombinants, it is not feasible to simply delete the Elb 3′ ss, as this overlaps with sequences required for pIX expression. The most straightforward approach has been to build upon what has been successful. The expression vector shown in Fig. 3a, for example, has reproducibly generated mRNAs of the expected structure, as well as high yields of several different proteins for which expression has been attempted.

Generation of helper-dependent Ad recombinants requires selection, and as described in Sect. 4 this has involved using the ability of SV40 TAg to complement the growth of Ad on otherwise nonpermissive CV1 cells. *Ad* recombinants containing either the SV40 early region per se or this region positioned downstream of either the herpes simplex virus (HSV) tk gene or py early region have been used to express TAg, tk, and the three py tumor antigens (THUMMEL et al. 1983; YAMADA et al. 1985; MANSOUR et al. 1985, 1986). Initially this approach was compromised by the excessive aberrant splicing events that gave rise to nonfunctional TAg-containing messages. More recently the positioning of genes so that they are fused to tripartite leader sequences has been found to eliminate these incorrect processing events (MANSOUR et al. 1986).

Probably the most significant contribution that helper-dependent Ad recombinants have made for maximizing heterologous gene expression is the identification of a sequence in the intron between the first and second leaders that effects a five- to ten-fold increase in transcription (MANSOUR et al. 1986). The identification of this sequence reconciled some previous observations, i.e., the sometimes comparable levels of expression achieved using helper-dependent or helper-independent Ad recombinants even though the former represented only a fraction of the viral population. Thus, with a helper-dependent recombinant expressing tk, only 25% of the population contained the tk gene, yet expression was as efficient as the late structural proteins (YAMADA et al. 1985), which was the greatest efficiency obtained with any helper-independent recombinant. Additionally, with a pymT-expressing Ad recombinant, which represented only 5% of the viral population, pymT levels were observed that approximated those reported using a helper-independent vector (MANSOUR et al. 1985; DAVIDSON and HASSELL 1987). The L1–L2 intronic sequence may not be the only feature accounting for these observations; however it certainly produces a dramatic increase in transcription. More recently this intronic sequence has been adapted for use in an expression cassette in a helper-dependent recombinant, and a 15-fold increase in expression levels has been reported (MASON et al. 1988).

Heterologous proteins expressed at high levels in Ad recombinants include the SV40 TAg; py small T, mT, and large T; tk; DHFR; ribonucleotide reductase; CAT; and factor IX (SOLNICK 1981; GLUZMAN et al. 1982; THUMMEL et al. 1983; BERKNER and SHARP 1984, 1985; YAMADA et al. 1985; DAVIS et al. 1985; MANSOUR et al. 1985, 1986; MASSIE et al. 1986; BERKNER et al. 1987; SCHAFFHAUSEN et al. 1987; DAVIDSON and HASSELL 1987). The highest levels demonstrated to date yield approximately 100–200 μg protein from a confluent 100 mm plate of 293 cells. Where appropriate, expression has included efficient protein secretion, phosphorylation, glycosylation, or carboxylation. In most of these studies the biological activity has been assayed and the heterologous protein has been demonstrated to be qualitatively authentic. In all studies to date, only one product per recombinant has been expressed. In producing ribonucleotide reductase, a heterodimer, an Ad recombinant expressing one subunit was used to infect a 293 cell line stably transformed with the other subunit, and the association of the subunits was demonstrated by consequent enzymatic activity (Huang et al. 1988). This approach is limited in producing abundant levels of both subunits, however, due to the Ad-induced inhibition of cellular translation at late times, when expression of the virus-encoded subunit would be highest. Two other approaches by which abundant expression of multiple subunits could be achieved are by coinfecting with Ad recombinants that encode individual peptides and by using an Ad recombinant expressing a polycistronic message.

6 Using Ad Vectors for Vaccine Production

The safe and efficacious administration of Ad4 and Ad7 serotypes to military recruits for the prevention of acute respiratory diseases has provided an impetus for developing Ad vectors for the production of vaccines. Adminstration of Ad vaccines is particularly convenient as they can be given orally in enteric-coated tablets, resulting in an asymptomatic infection that ultimately provides immunity against Ad respiratory disease. Serotypes Ad4, Ad7, and Ad5 have all been adapted as potential vaccine vectors.

The most intense effort in recombinant Ad vaccine development has been in expressing the hepatitis B surface envelope proteins, owing to the lack of an in vitro system for propagating hepatitis virus. Sequences encoding HBsAG, the major envelope protein, have either been substituted for the Ela or E3 region, or they have been inserted into the Ad genome in E3 or between the E4 region and the right ITR. In all cases where HBsAG expression was observed the protein was glycosylated, secreted, and assembled into 22 nm particles. Expression of the envelope-encoding protein sequences has been attempted using the endogenous HBV promoter or expression cassettes containing the MLP, or the hepatitis sequences have been positioned downstream of early promoters Ela or E3. The first case was the only instance in which translation was not detected (Saito et al. 1985, 1986). In these experiments DNA encoding 87% of the hepatitis B genome, including the promoter, was inserted at mu 93.3, immediately upstream of the E4 promoter. An Ad genome deleted in the E3 region was used to accommodate the additional hepatitis virus sequences, and the transcriptional orientation of the envelope gene was the same as the E4 region. Decreased synthesis of the E4 major transcript was observed, possibly due to read-through transcription from the envelope mRNA, but the virus was nonetheless viable. Late in infection abundant levels of hepatitis envelope mRNA appeared in the cytoplasm, illustrating the ability of heterologous genes embedded within the Ad genome to avoid the Ad-imposed block on cellular transcription and transport. However translation of envelope proteins was not observed, possibly due to the lack of a tripartite leader on the mRNA. As described in Sect. 5, HBsAG mRNA lacking a tripartite leader was inefficiently translated (DAVIS et al. 1985). Insertion of a cassette containing the MLP, a near-complete tripartite leader, and envelope protein sequences into the Ad genome either in E1 or between the E4 region and right ITR resulted in efficient translation of these sequences (MONLNAR-KIMBER et al. 1987, 1988; DAVID et al. 1985).

As mentioned HBV envelope-encoding sequences have also been positioned behind the E1a or E3 promoters (BALLAY et al. 1985; MORIN et al. 1987). The E3 recombinants express the envelope proteins at levels comparable to that obtained with vectors using the E1a promoter, i.e., approximately 1–3 μg per 10^6 cells. These vectors promote expression at early times, as expected. E1a-substituted Ad vectors expressed the envelope proteins both in the E1a-complementing 293 cell line and in VERO cells (BALLAY et al. 1985). However, since propagation of these viruses is limited to using a complementing cell line, the E3-substituted vectors may be of more value, since they can be produced in a broader range of cell lines. This includes WI38 (MORIN et al. 1987), a cell line which has already been approved for vaccine production.

Ad vectors containing HBV envelope protein sequences have been effective in eliciting seroconversion to the envelope proteins in vivo. Both wt and E3-deleted Ad inoculated into hamsters intranasally were capable of replicating in the lungs (MORIN et al. 1987). Even though E3-deleted Ad titers dropped more quickly than wt titers, consistent with the putative role of the E3 region in immunosurveillance, an antibody response to each virus was observed. In the case of Ad/HBV recombinant virus, antibody production against HBV proteins was also detected. Replication-defective E1-substituted Ad vectors also generated antibodies to HBV envelope proteins in rabbits (BALLAY et al. 1985) and in hamsters (MONLAR-KIMBER et al. 1987). The latter study was done in both the presence and absence of helper virus. Both permutations elicited envelope protein seroconversion, but antibody production against Ad proteins was only observed when hamsters were coinfected with helper virus.

The potential for vaccine production using Ad vectors has also been explored using the human immunodeficiency virus (HIV) envelope protein (HUNG et al. 1988), the measles virus hemagglutinin (ALKHATIB and BRIEDIS 1988), and the herpes simplex virus (HSV) gB glycoprotein (JOHNSON et al. 1988). Expression of all three proteins in Ad vectors has been demonstrated and both glycosylation and correct subcellular localization of the HSV and measles viral antigens have been observed. The measles virus hemagglutinin sequences were expressed in a cassette containing the MLP, substituted for the E1 region. In addition to being expressed in 293 cells, the gene was also efficiently expressed in HeLa and VERO cells when very high multiplicities of infection were used. Construction of the Ad/HSV gB protein recombinant involved positioning HSV sequences behind the SV40 promoter and substituting this for the E3 region. Expression of this recombinant was observed in human cells and, to a lesser extent, in murine cells. This is the instance alluded to in Sect. 4, however, where the SV40 promoter was shown to be inactive in E3 and where transcription likely derived from other upstream sequences, such as the MLP or E3 promoter. Thus, if the E3 region is the site of insertion or substitution of a heterologous gene, there appears to be no reason to include a heterologous promoter; it is not required and may not even be used due to upstream interference.

In addition to adaptation of Ad recombinants for vaccine production, there also may be a potential for developing animal models using Ad vectors, especially since Ad replicates in mouse cells. For example, the Ad recombinant expressing the HSV gB protein was used to confirm the observation that the gB glycoprotein is not a murine CTL target (JOHNSON et al. 1988). Expansion of this approach using Ad vectors may clearly be of value.

7 Analysis of Transcriptional Elements Incorporated into Ad Vectors

The ability of nonviral transcriptional elements to function when embedded within the Ad genome, coupled with the retention of tissue-specific RNA synthesis and processing, described below, has made Ad vectors well-suited for analyzing tissue-specific transcriptional events. The particular advantages of using Ad include their broad host range and efficiency of infection, which in many cell lines is 100%. Such an efficiency allows a comparison of expression of the endogenous and exogenous genes, as well as the ability to compare different cell types. Both of these measurements are difficult to make with alternative modes of gene introduction, where only a subpopulation of cells takes up DNA and where the fraction of cells expressing a gene can differ among cell types. Another advantage to using Ad vectors is that with gene fusions where foreign sequences are inserted behind early Ad promoters, detection is possible within hours after infection. This is in contrast to transfection experiments where steady state analysis of RNA and/or protein typically is performed after 24–72 h, requiring that assumptions be made about the mRNA and/or protein stability. Moreover, because Ad expression is so efficient, transcription rates rather than steady state values can be determined.

One application illustrating several of these advantages is an analysis of expression of the HIV-1 LTR and Rous sarcoma virus (RSV) LTR, which were each fused to the CAT gene and substituted in the E1 region of an Ad recombinant (RICE and MATHEWS 1988a, b). Combined analysis of nascent protein and RNA synthesis in HeLA cells which lack or contain the tat gene product was used to argue that transactivation of the HIV-1 LTR by tat operates at the level of transcription initiation rather than post-transcriptionally. The

RSV LTR was not transactivated by tat. However, both the RSV LTR and HIV-1 LTR were transactivated by E1a, specifically by the 13s gene product, in coinfection experiments where a helper virus provided the Ela gene. This Ela-mediated transactivation was observed both in HeLa cells and in two human leukemic T cell lines. Notably, in the absence of Ela, expression of the HIV-1 LTR-CAT gene was not observed in T cells, suggesting that an Ad vector expressing the HIV-1 LTR may provide diagnostic potential for detecting transactivation of this sequence in human lymphocytes.

The susceptibility of T cells to Ad infection, described above, is of interest since previously there were reports that cultured lymphocytes were highly refractory to Ad infection. More recently, however, it has been observed that several established myeloma B and T cell lines, which represent different developmental stages, are infected efficiently (LAVERY et al. 1987). This observation has been exploited by using an Ad recombinant to investigate how a single gene, the Ig heavy chain μ gene, can selectively give rise to mRNAs encoding secreted (μ_s) and membrane (μ_m) proteins in myeloma cells (RUETHER et al. 1986). The E1b promoter plus part of the 5′ E1b gene was fused to a fragment encoding a major portion of a mouse Ig μ heavy chain constant region gene, which included the alternative pA and splice signals that generate μ_s- or μ_m-mRNA. When a terminally differentiated human myeloma line was infected with the Ad recombinant, steady state accumulation of only the μ_s-mRNA was detected, appropriately. In contrast, no discrimination between the two messages was made in an epithelial-like line (HeLa), as both mRNAs were observed in approximately equal abundance. Analysis of nascent transcripts in myeloma cells indicated that correct polyadenylation was not due to transcriptional termination between the μ_s and μ_m sequences. Additionally, inserting the 3′ μ sequences behind the E1b promoter, thereby segregating them from the Ig promoter and enhancer, demonstrated that differential μ_s- and μ_m-mRNA expression is independent of these two elements.

Ad recombinants have been used to demonstrate tissue specific promoter activity in differentiated cells by fusing an albumin, Ig, or β-globin promoter to a promoterless E1b gene followed by infection of the appropriate cell lines with the recombinant viruses. The Ig promoter was active in a myeloma cell line while the albumin promoter was not, and only the albumin promoter functioned in a hepatoma cell line (FRIEDMAN et al. 1986). The activity of the endogenous albumin promoter in primary mouse cells declined rapidly after initial plating (i.e. within 24 h), and this promoter extinction was paralleled by the albumin promoter introduced in an Ad vector. Thus transcriptional factors responsible for tissue specific expression are available to promoters embedded within an Ad genome. Transcription specificity was observed both in cells containing low template copy numbers (i.e., using a defective, nonreplicating Ad recombinant containing the ablumin promoter) and in cells containing increased promoter copy numbers obtained by coinfecting the defective Ad recombinant with helper virus (FRIEDMAN et al. 1986; BABISS et al. 1987). When undifferentiated cells (i.e. HeLa) were infected with Ad recombinants containing the albumin or β-globin promoter both were active, regardless of the promoter copy number and despite the fact that the respective endogenous promoters were inactive. Promoter discrimination in differentiated cells versus promiscuity in undifferentiated cells clearly raises a number of questions. These would include whether tissue specificity arises from transcriptional activators or repressors and whether transcriptional factors are equally available to exogenous and endogenous promoters or are in some way selectively sequestered.

Although tissue specific expression of exogenously introduced promoters was observed, the level of mRNA synthesis was far less than that of the endogenous promoter (FRIEDMAN et al. 1986). Inclusion of the E1a enhancer upstream of the albumin promoter and coinfection with helper virus providing E1A stimulated transcription at least 100-fold, to a level surpassing that of the endogenous albumin promotor (BABISS et al. 1986). The mechanism explaining increased transcription is not clear. Possibly the E1a enhancer substitutes for another regulatory sequence in the albumin promoter, not present on this particular construct. Also not clear is the pool size of transcription factors and, as mentioned above, their availability to exogenously introduced promoters; the E1a enhancer and E1a, interestingly, had no affect on endogenous promoter activity. Although these studies are obviously limited at this point in providing a mechanistic interpretation of transcription specificity, they should serve as a valuable in vivo framework for in vitro analysis of transcription factor interaction with tissue specific promoters.

8 Establishing Stably Transduced Cell Lines Using Ad Vectors

Multiple selectable markers, including genes encoding tk, DHFR, and resistance to neomycin, as well as the entire SV40 early region, have been inserted into the Ad genome and used to generate stably transduced cell lines. Permutations have included incorporating the endogenous promoter of the gene, or fusing the coding sequences to the Ad2 MLP or the SV40 early promoter. Substitutions have been made in the E1 region or the E3 region using an Ad recombinant deleted in E1, or in both the E1 and E3 regions. Thus, a variety of selectable markers can be used in lieu of E1-mediated transformation to establish cell lines carrying stably integrated recombinant genomes (VAN DOREN et al. 1984). As with wt viruses, recombinants have also been found to be integrated in a structure colinear with that of the input viral genome (VAN DOREN and GLUZMAN 1984; GHOSH-CHOUDHURY and GRAHAM 1987; VAN DOREN et al. 1984; KARLSSON et al. 1985, 1986), with most of the genome represented in the integrant. Multiple copies, tandemly aligned and containing apparent right end-left end junctions, have also been reported (VAN DOREN et al. 1984). These patterns of integration are consistent with the fact that the viral vectors retain the E2 region which may be involved in integration events (GRAHAM 1984b).

Foreign genes positioned in either orientation in E1 or E3 have been used to generate transductants. With substitution within the E1 region, surrounding Ad sequences probably do not affect expression levels as dramatically as with substitution of the E3 region. Genes inserted in the left-to-right transcriptional direction in E3 may be affected by upstream promoters, as discussed in Sect. 4. Sequences inserted in the opposite orientation in E3 oppose MLP-derived synthesis and may be affected by transcription from the upstream E4 region. Critical evaluation of position and orientation effects on transcription patterns and, ultimately, on transduction frequencies have not been made, but this information may be of value for determining strategies for improving the frequency of viral transduction.

Ad recombinants containing cellular DHFR or tk genes have been used to infect recipient cell lines lacking endogenous DHFR (BERKNER and SHARP 1984; GHOSH-CHOUDHURY and GRAHAM 1987) or tk (HAJ-AHMAD and GRAHAM 1986) activities, generating stably transduced lines. In the case of the tk gene under the control of the cellular promoter, activity has been demonstrated as well during lytic infection. This is in contrast to the inability of the HBsAG gene, transcribed from its own promoter, to be translated in lytic infections (SAITO et al. 1985; Sect. 6 above). Whether this reflects a difference in the constructions, e.g. in mRNA stability or in the dependence of these individual mRNAs on the tripartite leader sequences, or is a consequence of the differences in cell types used or the detection levels of the two assays is unclear.

The dominant selectable neomycin-resistant gene has been used in Ad recombinants to transduce a variety of cell types, including established rat and simian cells (VAN DOREN et al. 1984), human fibroblasts, and human and mouse hematopoietic cells (KARLSSON et al. 1985, 1986). It also has been used in combination with a gene containing an endogenous, tissue specific promoter to demonstrate tissue specific expression. Ad recombinants were substituted in E3 by neomycin-encoding sequences and in E1 by a genomic human β-globin gene or a genomic human hybrid γ–β globin gene (KARLSSON et al. 1986). When these recombinants were used to infect a human erythroleukemia cell, expression of only the hybrid gene containing the γ-globin promoter was observed in transformants, as appropriate. The genes substituted for the E1 and E3 region each appeared to be under separate regulatory control, therefore offering an approach by which genes of interest and selectable markers can both be introduced into a single viral vector and stably integrated into DNA. An alternative for this might be provided by using polycistronic genes, which can be efficiently expressed in Ad-infected cells, as described in Sect. 5.

Substitution of E1 by the SV40 early region has been used to efficiently transform primary human fibroblasts (VAN DOREN and GLUZMAN 1984). SV40 can transform a variety of cell types which are normally refractory to culturing, such as human skeletal muscle cells and macrophages (MIRANDA et al. 1983; NAGATA et al. 1983). However, using SV40 as a viral vector can be problematic, as there are difficulties with recombination. Combining the advantages of Ad vectors with the ability of SV40 to transform different cell types therefore represents a valuable strategy for expanding the repertoire of transformed cell lines.

Although the frequency of Ad transduction is high (e.g., see VAN DOREN et al. 1984; KARLSSON et al. 1985; HAJ-AHMAD and GRAHAM 1986), it is still sufficiently less than that of retroviral vectors, limiting its utility in transforming stem cells of hematopoietic or bone marrow origin. However, by contrast with retroviral vectors, it offers the advantages that promoters, splicing and pA signals incorporated into the virus do not negatively interfere with viral propagation, and that multiple transcriptional units can be efficiently expressed from the same genome. Thus continued development of Ad recombinants, such as improving the expression of the selectable marker by incorporating the LTR into its transcription unit (GORMAN et al. 1983) or by possibly inserting the gene into another part of the genome, is certainly of interest.

9 Conclusions

The versatility and proven efficacy of Ad vectors will undoubtedly result in an expansion in the use of this system. There are clear areas for growth, such as developing second generation helper cell lines that express multiple Ad early regions or improving the vectors as additional insights are made about the biology of this virus. There is also untapped potential in a variety of areas, e.g., in exploiting the transduction efficiency of Ad by adapting Ad vectors for expression library screening, or in using Ad deletion recombinants as functional traps for isolating cellular proteins with Ad-like activities. One of the largest areas for growth involves adapting the advantages of other viral systems; the use of Ad/SV40 recombinants to broaden cell transformability as described above is an excellent example of the synergistic potential that might be provided by incorporating multiple viral features.

Acknowledgements. I would like to thank Donna Prunkard, Kent Beech, Julie Holly, and Deb Sawislak for their help during the generation of this manuscript, and Margo Rogers for its careful preparation. I am grateful to Kurt Runge and Sharon Busby for their editorial comments.

References

Aküsjarvi G, Svensson C, Nygård O (1987) A mechanism by which adenovirus virus-associated RNA_I controls translation in a transient expression assay. Mol Cell Biol 7: 549–551

Alkhatib G, Briedis DJ (1988) High-level eucaryotic in vivo expression of biologically active measles virus hemagglutinin by using an adenovirus type 5 helper-free vector system. J Virol 62: 2718–2727

Anderson CW, Lewis JB, Baum PR, Gesteland RF (1976) Simian virus 40-specific polypeptides in $Ad2^+ND1$- and $Ad2^+ND4$-infected cells. J Virol 18: 685–692

Andersson M, Pääbo S, Nilsson T, Peterson PA (1985) Impaired intracellular transport of class I MHC antigens as a possible means for adenoviruses to evade immune surveillance. Cell 43: 215–222
Babich A, Feldman LT, Nevins JR, Darnell JE Jr Weinberger C (1983) Effect of adenovirus on metabolism of specific host mRNAs: transport control and specific translational discrimination. Mol Cell Biol 3: 1212–1221
Babiss LE, Ginsberg HS (1984) Adenovirus type 5 early region 1b gene product is required for efficient shutoff of host protein synthesis. J Virol 50: 202–212
Babiss LE, Ginsberg HS, Darnell JE Jr (1985) Adenovirus E1B proteins are required for accumulation of late viral mRNA and for effects on cellular mRNA translation and transport. Mol Cell Biol 5: 2552–2558
Babiss LE, Friedman JM, Darnell JE Jr (1986) Cellular promoters incorporated into the adenovirus genome: effects of viral regulatory elements on transcription rates and cell specificity of albumin and β-globin promoters. Mol Cell Biol 6: 3798–3806
Babiss LE, Friedman JM, Darnell JE Jr (1987) Cellular promoters incorporated into the adenovirus genome. Effect of viral DNA replication on endogenous and exogenous gene transcription. J Mol Biol 193: 643–650
Ballay A, Levrero M, Buendia MA, Tiollais P, Pericaudet M (1985) In vitro and in vivo synthesis of the hepatitis B virus surface antigen and of the receptor for polymerized human serum albumin from recombinant human adenoviruses. EMBO J 4: 3861–3865
Barget SM, Moore C, Sharp P (1977) Spliced RNA segments at the 5′ terminus of adenovirus 2 late mRNA. Proc Natl Acad Sci USA 74: 3171–3175
Berk AJ (1986) Adenovirus promoters and E1A transactivation. Annu Rev Genet 20: 45–79
Berkner KL (1988) Development of adenovirus vectors for the expression of heterologous genes. Biotechniques 6: 616–629
Berkner KL, Sharp PA (1983) Generation of adenovirus by transfection of plasmids. Nucleic Acids Res 11: 6003–6020
Berkner KL, Sharp PA (1984) Expression of dihydrofolate reductase and of the adjacent E1b region in an Ad5-dihydrofolate reductase recombinant virus. Nucleic Acids Res 12: 1925–1941
Berkner KL, Sharp PA (1985) Effect of the tripartite leader on synthesis of a non-viral protein in an adenovirus 5 recombinant. Nucleic Acids Res 13: 841–857
Berkner KL, Schaffhausen BS, Roberts TM, Sharp PA (1987) Abundant expression of polyomavirus middle T antigen and dihydrofolate reductase in an adenovirus recombinant. J Virol 61: 1213–1220
Berkner KL, Boel E, Prunkard D (1988) Efficiency of translation of polycistronic messages in uninfected and adenovirus-infected cells. In: Gluzman Y, Hughes SH (eds) Viral vectors. Cold Spring Harbor Laboratory, Cold Spring Harbor NY, pp 56–61
Bridge E, Ketner G (1989) Redundant control of adenovirus late gene expression by early region 4. J Virol 63: 631–638
Burgert HG, Kvist S (1985) An adenovirus type 2 glycoprotein blocks cell surface expression of human histocompatibility class I antigens. Cell 41: 987–997
Burgert HG, Maryanski JL, Kvist S (1987) "E3/19K" Protein of adenovirus type 2 inhibits lysis of cytolytic T lymphocytes by blocking cell-surface expression of histocompatibility class I antigens. Proc Natl Acad Sci USA 84: 1356–1360
Castrillo JL, Carrasco L (1987) Adenovirus late protein synthesis is resistant to the inhibition of translation induced by poliovirus. J Biol Chem 262: 7328–7334
Chatterjee PK, Flint SJ (1987) Adenovirus type 2 endopeptidase: an unusual phosphoprotein enzyme matured by autocatalysis. Proc Natl Acad Sci USA 84: 714–718
Cherney CS, Wilhelm YM (1979) Differential translation in normal and adenovirus type 5-infected human cells and cell-free systems. J Virol 30: 533–542
Chinnadurai G, Chinnadurai S, Brussa J (1979) Physical mapping of a large-plaque mutation of adenovirus type 2. J Virol 32: 623–628
Chow LT, Roberts JM, Lewis JB, Broker TR (1977) A map of cytoplasmic RNA transcripts from lytic adenovirus type 2, determined by electron microscopy of RNA-DNA hybrids. Cell 11: 819–836
Colby WW, Shenk T (1981) Adenovirus type 5 virions can be assembled in vivo in the absence of detectable polypeptide IX. J Virol 39: 977–980
Cutt JR, Shenk T, Hearing P (1987) Analysis of adenovirus early region 4-encoded polypeptides synthesized in productively infected cells. J Virol 61: 543–552
Davidson D, Hassell JA (1987) Overproduction of polyomavirus middle T antigen in mammalian cells through the use of an adenovirus vector. J Virol 61: 1226–1239

Davis AR, Kostek B, Mason BB, Hsiao CL, Morin J, Dheer SK, Hung PP (1985) Expression of hepatitis B surface antigen with a recombinant adenovirus. Proc Natl Acad Sci USA 82: 7560–7564
Doerfler W (ed) (1983–1984) The molecular biology of adenoviruses, vols 1–3. Springer, Berlin Heidelberg New York (Current topics in microbioloy and immunology, vols 109–111)
Dolph PJ, Racaniello V, Villamarin A, Palladino F, Schneider RJ (1988) The adenovirus tripartite leader may eliminate the requirement for cap-binding protein complex during translation initiation. J Virol 62: 2059–2066
Falck-Pedersen E, Logan J, Shenk T, Darnell JE Jr (1985) Transcription termination within the E1A gene of adenovirus induced by insertion of the mouse β-major globin terminator element. Cell 40: 897–905
Falgout B, Ketner G (1987) Adenovirus early region 4 is required for efficient virus particle assembly. J Virol 61: 3759–3768
Friedman JM, Babiss LE, Clayton DF, Darnell JE Jr (1986) Cellular promoters incorporated into the adenovirus genome: cell specificity of albumin and immunoglobulin expression. Mol Cell Biol 6: 3791–3797
Gaynor RB, Hillman D, Berk AJ (1984) Adenovirus early region 1A protein activates transcription of a nonviral gene introduced into mammalian cells by infection or transfection. Proc Natl Acad Sci USA 81: 1193–1197
Ghosh-Choudhury G, Graham FL (1987) Stable transfer of a mouse dihydrofolate reductase gene into a deficient cell line using human adenovirus vector. Biochem Biophys Res Commun 147: 964–973
Ghosh-Choudhury G, Haj-Ahmed Y, Graham FL (1987) Protein IX, a minor component of the human adenovirus capsid, is essential for the packaging of full length genomes. EMBO J 6: 1733–1739
Ginsberg HS (1984) The adenoviruses. Plenum, New York
Gluzman Y, Reichl H, Solnick D (1982) Helper-free adenovirus type-5 vectors. In: Gluzman Y (ed) Eukaryotic viral vectors Cold Spring Harbor Laboratory, New York, pp 187–192
Gooding LR, Elmore LW, Tollefson AE, Brady HA, Wold WSM (1988) A 14,700 mw protein from the E3 region of adenovirus inhibits cytolysis by tumor necrosis factor. Cell 53: 341–346
Gorman C, Padmanabhan R, Howard BH (1983) High efficiency DNA-mediated transformation of primate cells. Science 221: 551–552
Graham FL (1984a) Covalently closed circles of human adenovirus DNA are infectious. EMBO J 3: 2917–2922
Graham FL (1984b) Transformation by and oncogenicity of human adenoviruses. In: Ginsberg HS (ed) The adenoviruses Plenum, New York, pp 339–398
Graham FL (1987) Growth of 293 cells in suspension culture. J Gen Virol 68: 937–940
Graham FL, Smiley J, Russell WC, Nairn R (1977) Characteristics of a human cell line transformed by DNA from human adenovirus 5. J Gen Virol 36: 59–72
Graham FL, Rowe DT, McKinnon R, Bacchetti S, Ruben M, Branton PE (1984) Transformation by human adenoviruses. J Cell Physiol [suppl] 3: 151–163
Grand RJA (1987) The structure and functions of the adenovirus early region 1 proteins. Biochem J 241: 25–38
Haj-Ahmad Y, Graham FL (1986) Development of a helper-independent human adenovirus vector and its use in the transfer of the herpes simplex virus thymidine kinase gene. J Virol 57: 267–274
Halbert DN, Cutt JR, Shenk T (1985) Adenovirus early region 4 encodes functions required for efficient DNA replication, late gene expression, and host cell shutoff. J Virol 56: 250–257
Hanahan D, Gluzman Y (1984) Rescue of functional replication origins from embedded configurations in a plasmid carrying the adenovirus genome. Mol Cell Biol 4: 302–309
Hay RT (1985a) The origin of adenovirus DNA replication: minimal DNA sequence requirement in vivo EMBO J 4: 421–426
Hay RT (1985b) The origin of adenovirus DNA replication: role of the nuclear factor I binding site in vivo J Mol Biol 186: 129–136
Hay RT, McDougall IM (1986) Viable viruses with deletions in the left inverted terminal repeat define the adenovirus origin of DNA replication. J Gen Virol 67: 321–332
Hay RT, Stow ND, McDougall IM (1984) Replication of adenovirus mini-chromosomes. J Mol Biol 175: 493–510
Hearing P, Shenk T (1983) The adenovirus type 5E1A transcriptional control region contains a duplicated enhancer element. Cell 33: 695–703

Hearing P, Shenk T (1985) Sequence-independent autoregulation of the adenovirus type 5 E1A transcription unit. Mol Cell Biol 5: 3214–3221
Hearing P, Samulski RJ, Wishart WL, Shenk T (1987) Identification of a repeated sequence element required for efficient encapsidation of the adenovirus type 5 chromosome. J Virol 61: 2555–2558
Hirt B (1967) Selective extraction of polyoma DNA from infected mouse cell cultures. J Mol Biol 26: 365–369
Huang A, Jacobi G, Haj-Ahmad Y, Bacchetti S (1988) Expression of the HSV-2 ribonucleotide reductase subunits in adenovirus vectors or stably transformed cells: restoration of enzymatic activity by reassociation of enzyme subunits in the absence of other HSV proteins. Virology 163: 462–470
Hung PP, Morin JE, Lubeck MD, Barton JE, Molnar-Kimber KL, Mason BB, Dheer SK, Jarocki-Witek V, Kostek B, Zandle G, Conley AJ, Davis AR (1988) Recombinant adenovirus as a vehicle for the HBV surface antigen or HIV envelope protein genes. In: Human retroviruses, cancer and AIDS: approaches to prevention and therapy. Liss, New York, pp 349–361
Jang SK, Kräusslich HG, Nicklin MJH, Duke GM, Palmenberg AC, Wimmer E (1988) A segment of the 5′ nontranslated region of encephalomyocarditis virus RNA directs internal entry of ribosomes during in vitro translation. J Virol 62: 2636–2643
Johnson DC, Ghosh-Choudhury G, Smiley JR, Falls L, Graham FL (1988) Abundant expression of herpes simplex virus glycoprotein gB using an adenovirus vector. Virology 164: 1–14
Jones N, Shenk T (1978) Isolation of deletion and substitution mutants of adenovirus type 5. Cell 13: 181–188
Jones N, Shenk T (1979) Isolation of adenovirus type 5 host range deletion mutants defective for transformation of rat embryo cells. Cell 17: 683–589
Kämpe O, Bellgrau D, Hammerling U, Lind P, Pääbo S, Severinsson L, Peterson PA (1983) Complex formation of class 1 transplantation antigens and a viral glycoprotein. J Biol Chem 258: 19594–10598
Karlsson S, Humphries RK, Gluzman Y, Nienhuis AW (1985) Transfer of genes into hematopoietic cells using recombinant DNA viruses. Proc Natl Acad Sci USA 82: 158–162
Karlsson S, Van Doren K, Schweiger SG, Nienhuis AW, Gluzman Y (1986) Stable gene transfer and tissue-specific expression of a human globin gene using adenoviral vectors. EMBO J 5: 2377–2385
Katze MG, DeCorato D, Safer B, Galabru J, Hovanessian AG (1987) Adenovirus VAI RNA complexes with the 68,000 M_r protein kinase to regulate its autophosphorylation and activity. EMBO J 6: 689–697
Kaufman RJ (1985) Identification of the components necessary for adenovirus translation control and their utilization in cDNA expression vectors. Proc Natl Acad Sci USA 82: 689–693
Kelly TJ Jr (1984) Adenovirus DNA replication. In: Ginsberg HS (ed) The adenoviruses. Plenum, New York, pp 271–308
Kelly TJ, Lewis AM (1973) Use of nondefective adenovirus-simian virus 40 hybrids for mapping the SV40 genome. J Virol 12: 643–652
Khalili K, Weinmann R. (1984) Shut-off of actin biosynthesis in adenovirus serotype-2-infected cells. J Mol Biol 175: 453–468
Kvist S, Oestberg L, Persson H, Philipsson L, Peterson, P.A (1978) Molecular association between transplantation antigens and a cell surface antigen in an adenovirus-transformed cell line. Proc Natl Acad Sci USA 75: 5674–5678
Lai Fatt RB, Mak S (1982) Mapping of an adenovirus function involved in the inhibition of DNA degradation. J Virol 42: 969–977
Lavery D, Fu SM, Lufkin T, Chen-Kiang S (1987) Productive infection of cultured human lymphoid cells by adenovirus. J Virol 61: 1466–1472
Levine AJ (1984) The adenovirus early proteins. Curr Top Microbiol Immunol 110: 143–167
Logan J, Shenk T (1984) Adenovirus tripartite leader sequence enhances translation of mRNAs late after infection. Proc Natl Acad Sci USA 81: 3655–3659
Mansour SL, Grodzicker T, Tjian R (1985) An adenovirus vector system used to express polyoma virus tumor antigens. Proc Natl Acad Sci USA 82: 1359–1363
Mansour SL, Grodzicker T, Tjian R (1986) Downstream sequences affect transcription initiation from the adenovirus major late promoter. Mol Cell Biol 6: 2684–2694
Mason BB, Morin JE, Davis AR, Conley AJ, Lubeck MD, Molnar-Kimber KL, Chengalvala M, Bhat B, Dheer S, Hum WT, Hung PP (1988) Development of recombinant adenovirus as a vaccine. In: Gluzman Y, Hughes SH (eds) Viral vectors. Cold Spring Harbor Laboratory, Cold Spring Harbor N Y, pp 51–55

Massie B, Gluzman Y, Hassell JA (1986) Construction of a helper-free recombinant adenovirus that expresses polyomavirus large T antigen. Mol Cell Biol 6: 2872–2883
McGrory WJ, Bautista DS, Graham FL (1988) A simple technique for the rescue of early region I mutations into infectious human adenovirus type 5. Virology 163: 614–617
Miller BW, Williams J (1987) Cellular transformation by adenovirus type 5 is influenced by the viral DNA polymerase. J Virol 61: 3630–3634
Miranda AF, Babiss LE, Fisher PB (1983) Transformation of human skeletal muscle cells by simian virus 40. Proc Natl Acad Sci USA 80: 6581–6585
Molnar-Kimber KL, Davis AR, Jarocki-Witek V, Lubek MD, Vernon SK, Conley AJ, Hung PP (1987) Characterization and assembly of hepatitis B envelope proteins expressed by recombinant adenovirus. UCLA Symp Mol Cell Biol 70: 173–187
Molnar-Kimber KL, Jarocki-Witek V, Dheer SK, Vernon SK, Conley AJ, Davis AR, Hung PP (1988) Distinctive properties of the hepatitis B virus envelope proteins. J Virol 62: 407–416
Moore MA, Shenk T (1988) The adenovirus tripartite leader sequence can alter nuclear and cytoplasmic metabolism of a non-adenovirus mRNA within infected cells. Nucleic Acids Res 16: 2247–2262
Moore M, Schaack J, Baim SB, Morimoto RI, Shenk T (1987) Induced heat shock mRNAs escape the nucleocytoplasmic transport block in adenovirus-infected HeLa cells. Mol Cell Biol 7: 4505–4512
Morin JE, Lubeck MD, Barton JE, Conley AJ, Davis AR, Hung PP (1987) Recombinant adenovirus induces antibody response to hepatitis B virus surface antigen in hamsters. Proc Natl Acad Sci USA 84: 4626–4730
Munz PL, Young CSH (1987) The creation of adenovirus genomes with viable, stable, internal redundancies centered about the E2b region. Virology 158: 52–60
Nagata Y, Diamond B, Bloom BR (1983) The generation of human monocyte/macrophage cell lines. Nature 306: 597–599
Nevins JR (1987) Regulation of early adenovirus gene expression. Microbiol Rev 51: 419–430
Nevins JR, Winkler JJ (1980) Regulation of early adenovirus transcription: a protein product of early region 2 specifically represses region 4 transcription. Proc Natl Acad Sci USA 77: 1893–1897
O'Malley RP, Mariano TM, Siekierka J, Mathews MB (1986) A mechanism for the control of protein synthesis by adenovirus VA RNA_I. Cell 44: 391–400
Pääbo S, Weber F, Kämpe O, Schaffner W, Peterson PA (1983) Association between transplantation antigens and a viral membrane protein synthesized from a mammalian expression vector. Cell 33: 445–453
Pääbo S, Bhat BM, Wold WSM, Peterson PA (1987) A short sequence in the COOH-terminus makes an adenovirus membrane glycoprotein a resident of the endoplasmic reticulum. Cell 50: 311–317
Pelletier J, Sonenberg N (1988) Internal initiation of translation of eukaryotic mRNA directed by a sequence derived from poliovirus RNA. Nature 334: 320–325
Persson H, Jansson M, Philipson L (1980) Synthesis and genomic site for an adenovirus type 2 early glycoprotein. J Mol Biol 136: 375–394
Pilder S, Logan J, Shenk T (1984) Deletion of the gene encoding the adenovirus 5 early region 1B 21,000-molecular-weight polypeptide leads to degradation of viral and host cell DNA. J Virol 52: 664–671
Pilder S, Moore M, Logan J, Shenk T (1986) The adenovirus E1B–55k transforming polypeptide modulates transport or cytoplasmic stabilization of viral and host cell mRNAs. Mol Cell Biol 6: 470–476
Powell J, Berkner K, Lebo R, Adamson J (1986) Human erythroproietin gene: high level expression in stably transfected mammalian cells and chromosome localization. Proc Natl Acad Sci USA 83: 6465–6469
Rice AP, Mathews MB (1988a) Transcriptional but not translational regulation of HIV-1 by the *tat* gene product. Nature 332: 551–553
Rice AP, Mathews MB (1988b) Trans-activation of the human immunodeficiency virus long terminal repeat sequences, expressed in an adenovirus vector, by the adenovirus E1A 13S protein. Proc Natl Acad Sci USA 85: 4200–4204
Rowe WP, Huebner RJ, Gillmore LK, Parrott RH, Ward TG (1953) Isolation of a cytopathogenic agent from human adenoids undergoing spontaneous degeneration in tissue culture. Proc Soc Exp Biol Med 84: 570–573
Ruben M, Bacchetti S, Graham F (1983) Covalently closed circles of adenovirus 5 DNA. Nature 301: 172–174

Ruether JE, Maderious A, Lavery D, Logan J, Fu SM, Chen-Kiang S (1986) Cell-type-specific synthesis of murine immunoglobulin μ RNA from an adenovirus vector. Mol Cell Biol 6: 123–133
Saito I, Oya Y, Yamamoto K, Yuasa T, Shimojo H (1985) Construction of nondefective adenovirus type 5 bearing a 2.8-kilobase hepatitis B virus DNA near the right end of its genome. J Virol 54: 711–719
Saito I, Oya Y, Shimojo H (1986) Novel RNA family structure of hepatitis B virus expressed in human cells, using a helper-free adenovirus vector. J Virol 58: 554–560
Sandler AB, Ketner G (1989) Adenovirus early region 4 is essential for normal stability of late nuclear RNAs. J Virol 63: 624–630
Sarnow P, Hearing P, Anderson CW, Halbert DN, Shenk T, Levine AJ (1984) Adenovirus early region 1b 58,000-dalton tumor antigen is physically associated with an early region 4 25,000-dalton protein in productively infected cells. J Virol 49: 692–700
Schaffhausen BS, Bockus BJ, Berkner KL, Kaplan D, Roberts TM (1987) Characterization of middle T antigen expressed by using an adenovirus expression system. J Virol 61: 1221–1225
Signäs C, Katze MG, Persson H, Philipson L (1982) An adenovirus glycoprotein binds heavy chains of class 1 transplantation antigens from man and mouse. Nature 299: 175–178
Solnick D (1981) Construction of an adenovirus-SV40 recombinant producing SV40 T antigen from an adenovirus late promoter. Cell 24: 135–143
Stillman BW (1985) Biochemical and genetic analysis of adenovirus DNA replication in vitro In: Setlow JK, Hollaender A (eds) Genetic engineering: principles and methods. Plenum, New York, pp 1–27
Stow ND (1988) Cloning of a DNA fragment from the left-hand terminus of the adenovirus type 2 genome and its use in site-directed mutagenesis. J Virol 37: 171–180
Subramanian T, Kuppuswamy M, Gysbers J, Mak S, Chinnadurai G (1984) 19 kDa Tumor antigen coded by early region E1b of adenovirus 2 is required for efficient synthesis and for protection of viral DNA. J Biol Chem 259: 11777–11783
Sevensson C, Aküsjarvi G (1985) Adenovirus VA RNA_I mediates a translation stimulation which is not restricted to the viral mRNAs. EMBO J 4: 957–964
Thimmappaya B, Weinberger C, Scheider RJ, Shenk T (1982) Adenovirus VAI RNA is required for efficient translation of viral mRNAs at late times after infection. Cell 31: 543–551
Thummel C, Tjian R, Grodzicker T (1982) Construction of adenovirus expression vectors by site-directed in vivo recombination. J Mol Appl Genet 1: 435–446
Thummel C, Tjian R, Hu SH, Grodzicker T (1983) Translational control of SV40 T antigen expressed from the adenovirus late promoter. Cell 33: 455–464
Tollefson AE, Wold WSM (1988) Identification and gene mapping of a 14,700-molecular-weight protein encoded by region E3 of group C adenoviruses. J Virol 62: 33–39
Tooze J (ed) (1981) DNA tumor viruses: molecular biology of tumor viruses, 2nd edn. Cold Spring Harbor Laboratories, Cold Spring Harbor, NY (Cold Spring Harbor monograph ser, vol 10b)
Van Doren K, Gluzman Y (1984) Efficient transformation of human fibroblasts by adenovirus-simian virus 40 recombinants. Mol Cell Biol 4: 1653–1656
Van Doren K, Hanahan D, Gluzman Y (1984) Infection of eucaryotic cells by helper-independent recombinant adenoviruses: early region 1 is not obligatory for integration of viral DNA. J Virol 50: 606–614
Weinberg DH, Ketner G (1983) A cell line that supports the growth of a defective early region 4 deletion mutant of human adenovirus type 2. Proc Natl Acad Sci USA 80: 5383–5386
Weinberg DH, Ketner G (1986) Adenoviral early region 4 is required for efficient viral DNA replication and for late gene expression. J Virol 57: 833–838
White E, Grodzicker T, Stillman BW (1984) Mutations in the gene encoding the adenovirus early region 1B 19,000-molecular-weight tumor antigen cause the degradation of chromosomal DNA. J Virol 52: 410–419
Williams JF (1986) Adenovirus genetics. In: Doerfler W (ed) Adenovirus DNA: the viral genome and its expression. Nijhoff, The Hague, pp 247–309
Wold WSM, Cladaras C, Magie SC, Yacoub N (1984) Mapping a new gene that encodes an 11,600-molecular-weight protein in the E3 transcription unit of adenovirus 2. J Virol 52: 307–313
Yamada M, Lewis JA, Grodzicker T (1985) Overproduction of the protein product of a nonselected foreign gene carried by an adenovirus vector. Proc Natl Acad Sci USA 82: 3567–3571
Yoder SS, Berget SM (1986) Role of adenovirus type 2 early region 4 in the early-to-late switch during productive infection. J Virol 60: 779–781

Epstein-Barr Virus Based Expression Vectors

R. F. MARGOLSKEE

1 Introduction

Eukaryotic viral vectors have played an important role in the expression and analysis of many eukaryotic genes. The field is growing at a rapid pace and is by no means mature (for contrast see the reports from the first and second Banbury Conferences on Eukaryotic Viral Vectors, GLUZMAN 1982; GLUZMAN and HUGHES 1988). Although in theory it should be possible to develop a eukaryotic vector from any virus, practical considerations have led to concentration on a few well-characterized viruses, e.g., papovaviruses, adenoviruses,

Department of Neurosciences, Roche Institute of Molecular Biology, Roche Research Center, Nutley, NJ 10770, USA

Current Topics in Microbiology and Immunology Vol 158

herpesviruses, and retroviruses. As the understanding of the underlying biology of the virus' life cycle has matured, so too has the sophistication of specific virus based vectors. Frequently the development of virus vectors has been instrumental in dissecting the functional elements of the parent virus. Present-day eukaryotic vectors are a collection of eukaryotic elements from many different cellular and viral sources. Unfortunately, there are no simple rules for constructing the best eukaryotic vector, hence much of this field has developed from empirical trials. Furthermore, no single vector presently exists that provides a solution for all situations that require a eukaryotic vector.

Existing eukaryotic vectors can be categorized as freely replicating plasmids (episomal) or nonreplicating (integrating) vectors. Either type of vector may suffice for short-term, transient expression assays. In a typical transient assay, eukaryotic cells are transfected with vector DNA, then 1 to 3 days later expression of the vector encoded gene or cDNA is monitored. Transient assays are dependent upon entry and expression of vector encoded genes, but independent of vector integration. However, stable expression requires that the vector be maintained in the host cell in an unrearranged form for a matter of weeks to months. Replicating plasmid vectors have several clear advantages over integrating vectors in stable assays: (a) Stable transfection efficiency is usually much higher with plasmid vectors than with integrating vectors. (b) Cells transformed with a replicating plasmid are uniform in their expression of vector encoded genes; there is no variability derived from the integration locus or the integrant form. (c) Recovery of plasmid vectors is a much simpler and more efficient process than recovery of integrants. These combined advantages make eukaryotic plasmid vectors extremely useful in experiments utilizing stable expression, expression cloning, shuttle mutagenesis, or directed mutagenesis.

Nonreplicating vectors are manifold and include derivatives of papovaviruses, herpesviruses, adenoviruses, adeno-associated viruses, and retroviruses. Replicating plasmid vectors have been derived from bovine papillomavirus (BPV) and Epstein-Barr virus (EBV). Unfortunately, BPV vectors are not reliably maintained as stable plasmids (for review see MECSAS and SUGDEN 1987). BPV vectors all too often undergo integration, deletion, recombination, and rearrangement. BPV vector stability is affected by the host cell, the DNA insert, and the specific vector chosen. Futhermore, it has not been possible to predict the behavior of a given BPV construct. For this reason many workers have turned to EBV to develop a more reliable eukaryotic plasmid. EBV vectors, as with many modern vectors, utilize elements derived from many sources: bacterial (drug resistance genes, replication origins, and promoters); eukaryotic cellular and viral genes and their regulatory elements (replication origins, promoters, splice sites, and polyadenylation signals). It is the EBV replicon (EBV replication origin and EBNA-1) which confers upon these vectors their unique ability to be maintained as stable, freely replicating plasmids.

2 EBV Biology and Genetics

EBV is a human herpesvirus which commonly infects B cells to cause infectious mononucleosis (HENLE and HENLE 1972). Although mononucleosis is usually a self-limited lymphoproliferative disorder, in individuals with certain T cell deficiencies a fatal disease may ensue (PURTILLO et al. 1978). EBV has also been implicated in the pathogenesis of Burkitt's lymphoma, nasopharyngeal carcinoma (for review see EPSTEIN and ACHONG 1979; KLEIN 1979) and thymic carcinoma (LEYVRAZ et al. 1985). In immunosuppressed individuals (e.g., transplant and AIDS patients) there is a significant risk of developing EBV-associated lymphomas (BIRX et al. 1986; CLEARY et al. 1986).

2.1 Latent Replication: the *ori* P/EBNA-1 Replicon

EBV is capable of transforming human B cells in vitro to a continuously dividing, immortalized state (POPE et al. 1968; PATTENGALE et al. 1973; HENLE et al. 1987). Cells transformed by EBV maintain the complete virus genome as a covalently closed cirucular plasmid (NONOYAMA and PAGANO 1972; LINDAHL et al. 1976). Copy number varies from 1 to 800 per cell, but is characteristic of the given transformant (SUGDEN et al. 1979). The vast majority of in vitro infected B cells maintain the virus in a latent state (for review see BAICHWAL and SUGDEN 1988) wherein only a few viral genes are expressed. Latently infected B cells rarely give rise to lytic growth of EBV; depending upon the clonal isolate this may occur in less than one cell in a million (SUGDEN et al. 1979; SUGDEN 1984). In latently infected cells the virus genome replicates bidirectionally, approximately once per cell cycle in synchrony with the host DNA (NONOYAMA and PAGANO 1972; HAMPAR et al. 1974; ADAMS and LINDAHL 1975; GUSSANDER and ADAMS 1984; ADAMS 1987). In lytically infected cells EBV DNA replicates via a rolling circle intermediate to achieve a 1000-fold increase in copy number (MILLER 1985). The host range for EBV includes humans and a few other primates. EBV infects B cells and some nasopharyngeal epithelial cells which express the C3d complement receptor (FINGEROTH et al. 1984; FRADE et al. 1985). This is apparently the receptor for virus entry into these cells.

EBV virions contain a linear DNA molecule of 172 kilobases (kb) (DAMBAUGH et al. 1980; SKARE and STROMINGER 1980; BAER et al. 1984). After infection, the DNA molecule presumably circularizes via its terminal repeats and is retained as an episome in the cell's nucleus (LINDAHL et al. 1976; KINTNER and SUGDEN 1979). Plasmid replication of EBV derived replicons, and by inference EBV itself, has been shown to be dependent upon only two viral elements: the *cis* acting origin for plasmid replication (*ori* P), and the *trans*-acting EBV nuclear antigen (EBNA-1) (YATES et al. 1984, 1985; LUPTON and LEVINE 1985). The *Bam*HI restriction map of the B95.8 strain of EBV is shown in Fig. 1.

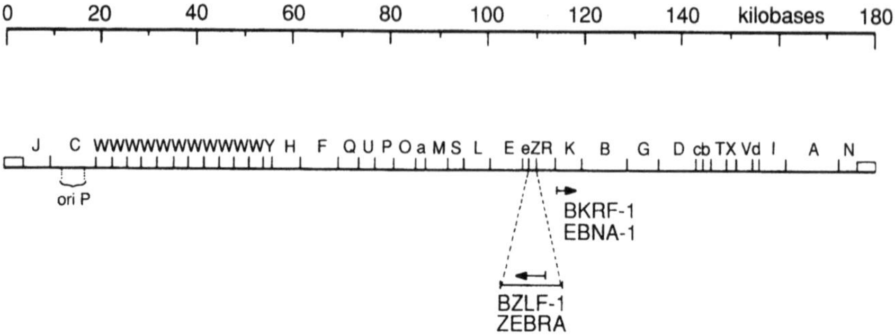

Fig. 1. *Bam*H1 restriction endonuclease map of the B95.8 strain of EBV. The genome is depicted as a 172 kb linear molecule, as it is found in the virion. Upon infection of B lymphoblasts the molecule circularizes via terminal repeats (*open boxes*). The *cis*-acting origin of plasmid replication (*ori* P) is located within the *Bam*H1 C fragment. The open reading frames encoding EBNA-1 and ZEBRA are indicated by the *arrows*. The designation BKRF1 indicates *Bam*H1 fragment rightward open reading frame 1; BZLF1 indicates *Bam*H1 Z fragment leftward open reading frame 1

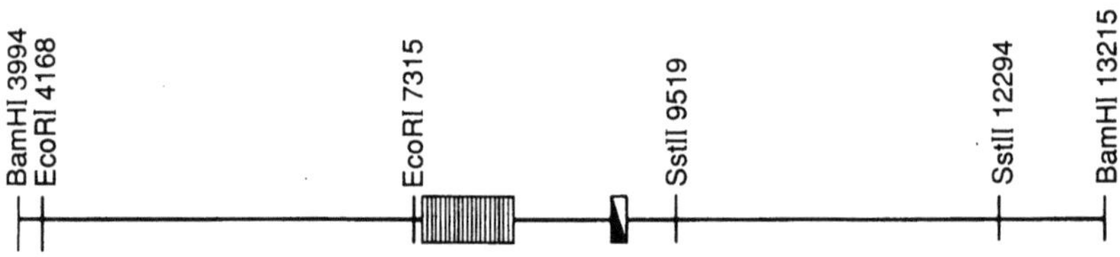

Fig. 2. Map of the *Bam*H1 C fragment of EBV. This region covers 9221 bp from positions 3994 to 13215 (nucleotide numbering is as in BAER et al. 1984). The EBV origin for plasmid replication (*ori* P) is within an 1800 bp segment of DNA (positions 7315 to 9106). *ori* P is comprised of two *cis*-acting elements: a family of 20 tandem repeats of a related 30 bp sequence (indicated by *vertically striped rectangle*) and a 65 bp dyad symmetry element (*half-shaded box*). Both the tandem repeats and the dyad element in *cis* are necessary and sufficient for EBNA-1 dependent plasmid replication

SUGDEN and coworkers (YATES et al. 1984) mapped *ori* P to a 1.8 kb region of the *Bam*HI C fragment of EBV DNA (see Fig. 2). The origin for plasmid replication is distinct from the origin for lytic replication, and *ori* P containing plasmids are not amplified in cells induced to undergo lytic production of EBV (YATES et al. 1984). Two distinct elements comprise *ori* P: (a) a family of similar 30 bp long A-T rich sequences repeated 20 times in tandem and (b) a 125 bp region containing a 65 bp dyad symmetry element comprised of sequences related to the tandem repeat units. The normal spacing between these two elements of *ori* P is about 900 bp (HUDSON et al. 1985). Deletion of either the family of repeats or the dyad symmetry element abolishes episomal replication (REISMAN et al. 1985; LUPTON and LEVINE 1985). Altering the spacing of these two elements by deletion or insertion of up to 1377 bp of extraneous DNA did not hinder *ori* P replication. Insertion of 4.6 kb of λ DNA did abolish plasmid replication (REISMAN et al. 1985). However, it was uncertain if this effect was due to the increased distance between elements or to something encoded within the λ DNA. Furthermore, altering the relative orientation of these two elements does not affect plasmid replication. Two dimensional gel electrophoresis indicates that DNA replication initiates at or near the dyad element within *ori* P, while the tandem repeats serve as a replication fork barrier (GAHN and SCHILDKRAUT 1989). Functional replication is maintained when only 7 to 8 of the 20 copies of the tandem repeats are present (WYSOKENSKI and YATES 1989; CHITTENDEN et al. 1989). The tandem repeats function as an EBNA-1 dependent enhancer of transcription in several cell lines including rodent cells non-permissive for *ori* P replication (REISMAN et al. 1985; LUPTON and LEVINE 1985; REISMAN and SUGDEN 1986; WYSOKENSKI and YATES 1989). EBNA-1 has been shown to bind in vitro to each copy of the tandem repeats and to the related sequences within the dyad symmetry element (RAWLINS et al. 1985). Substitution of the SV40 72-bp enhancer for the tandem repeat element of *ori* P does not restore plasmid replication (REISMAN and SUGDEN 1986). However, multiple tandem copies of the dyad element can substitute for the tandem family of repeats to provide near normal levels of replication (WYSOKENSKI and YATES 1989). This suggests a role for the EBNA-1 dependent enhancer in both transcription and replication, presumably via binding of EBNA-1. The tandem repeat enhancer is also required for EBNA-1 dependent enhancement of transcription of a promoter within the *Bam*HI C fragment of EBV DNA (LUPTON and LEVINE 1985; SUGDEN and WARREN 1989). This promoter is probably the one which is normally utilized for EBNA-1 transcription in latently infected cells (see below). As such, this suggests that EBNA-1 regulates its own transcription.

The only viral encoded *trans*-acting factor required for plasmid replication of EBV derived episomes is EBNA-1. Several nuclear antigens, including EBNA-1, are expressed in latently infected cells (HENLE et al. 1970; REEDMAN and KLEIN 1973; FRESEN et al. 1977; STRNAD et al. 1981; SUMMERS et al. 1982; DAMBAUGH et al. 1984; HENNESSY et al. 1985, 1986; HENNESSY and KIEFF 1985; KALLIN et al. 1986; DILLNER et al. 1986) but EBNA-1 is the only nuclear antigen in all EBV containing cells. EBNA-1 is a 65–80 kDa protein encoded by a 2 kb open reading frame in *Bam*HI K (the BKRF-1 open reading frame) (see Fig. 2) (HENNESSY and KIEFF 1983; FISCHER et al. 1984; HUDSON et al. 1985). Variability in the size of EBNA-1 arises from strain differences in the IR-3 repeat region of EBNA-1; this repeat region encodes over 230 resides of a glycine-alanine copolymer found within the NH_2 terminal portion of the protein (HENNESSY et al. 1983; HENNESSY and KIEFF 1983). The transcript for EBNA-1 is 3.7 kb in length and includes at least seven exons (SPECK and STROMINGER 1985) derived from several noncontiguous regions of EBV DNA (*Bam*HI fragments W, Y, U, E, and K). The unspliced precursor RNA for EBNA-1 must derive from over 100 kb of the viral genome (SPECK and STROMINGER 1985; BODESCOT et al. 1986; BODESLOT and PERRICAUDET 1986). The presumed promoter for this transcript is within *Bam*HI C (SPECK and STROMINGER 1987; BODESCOT et al. 1987).

Plasmid vectors containing both *ori* P and EBNA-1 replicate as autonomous plasmids in permissive cells (YATES et al. 1985; LUPTON and LEVINE 1985).

EBNA-1 may be supplied in *trans* by prior transformation of the host cell with EBV or with appropriate restriction fragments containing the EBNA-1 open reading frame (*Bam*HI K) linked to a promoter (Summers et al. 1982; Yates et al. 1985; Takada et al. 1986b; Hammarskjold et al. 1986). If the cell lacks EBNA-1, this function must be supplied by the vector. With *ori* P/EBNA-1 plasmids, large deletions and frameshift mutations within the coding region of EBNA-1 abolish replication (Lupton and Levine 1985).

2.2 Lytic Replication: *ori* lyt and ZEBRA

EBV is also capable of a lytic mode of replication which requires *cis* and *trans* elements distinct from *ori* P and EBNA-1. The *cis* element *ori* lyt (origin for lytic replication) has only recently been identified: the minimal lytic origin maps between bases 52623 and 53581 of EBV (Hammerschmidt and Sugden 1988). The minimal *ori* lyt element contains several arrays of inverted and direct repeats. One subelement of *ori* lyt is apparently a promoter/enhancer which can be functionally replaced by a promoter/enhancer from CMV (Hammerschmidt and Sugden 1988). "ZEBRA", the *trans*-acting factor required for induction of lytic replication, is the product of the BZLF-1 open reading frame within the *Bam*HI Z fragment (see Fig. 1) (Takada et al. 1986a; Countryman et al. 1987; Grogan et al. 1987). Depending upon the EBV strain, ZEBRA consists of a heterogeneous collection of polypeptides varying in size from 37 to 43 kDa.

Prior to the identification of ZEBRA and *ori* lyt, lytic induction of EBV was known to occur after exposure of some latently infected cells to several different chemicals or to superinfection with certain defective strains of EBV (Hampar et al. 1972; Gerber 1972; Tovy et al. 1978; zur Hausen et al. 1978; Luka et al. 1979; Takada 1984; Takada and zur Hausen 1984; Laux et al. 1988). Most commonly, the phorbol ester 12-0-tetradecanoylphorbol 13-acetate (TPA) has been used to induce the lytic cycle. This induction has a slow onset requiring 4–5 days for maximal lytic DNA replication.

ZEBRA induced lytic replication has a similar slow onset. ZEBRA is known to be a *trans*-activator of several EBV genes (Takada et al. 1986a; Chevallier-Greco et al. 1986). Among the genes activated by ZEBRA are at least three other EBV encoded *trans*-activators (Oguro et al. 1987). Although its exact mode of activation is presently not known, ZEBRA in the absence of any other EBV gene product can activate the promoter for the BSFL2 + BMLF1 trans-activator gene of EBV (Rooney et al. 1989). This suggests that the induction of the lytic cycle may depend upon a cascade of gene and/or protein activation extending over a period of days before the maximal effect is achieved. Among the various EBV encoded genes, only ZEBRA is capable of inducing the lytic cycle of replication. Even though ZEBRA induction of the lytic cycle is a delayed process, it is quite efficient: when plasmids containing an active promoter upstream of BZLF-1 are introduced into latently infected cells, up to 50% of the transfected cells are induced into the lytic cycle (Grogan et al. 1987).

Regulation of ZEBRA is normally quite stringent; in some latently infected cells only one cell in a million enters the lytic cycle (SUGDEN et al. 1979; SUGDEN 1984). In a defective genome of EBV (carried by the P3HR-1 cell line), rearrangement of the viral DNA juxtaposes BZLF-1 to a promoter which is expressed during latent infection (JENSON et al. 1987). This strain of EBV is proficient in lytic induction when it superinfects latent cells. In wild type EBV, induction of the lytic cycle is correlated with transcription of BZLF-1. It may be significant that in latent cells the primary transcript for EBNA-1 extends across BZLF-1 in the antisense direction. Perhaps this serves to suppress BZLF-1 transcription and/or translation of the ZEBRA protein.

3 EBV Based Eukaryotic Vectors

The wild type EBV genome exists as a 172 kb closed circular multicopy plasmid in the nucleus of latently infected cells. This genome would be too large for ready manipulation as a recombinant plasmid to shuttle between *E. coli* and mammalian cells. However, the *cis* and *trans* elements necessary for plasmid maintenance of EBV and EBV derived plasmids have been determined (see Sect. 2.1).

3.1 *ori* P Vectors

The first EBV based plasmid, pBam C, contained the 9.3 kb *Bam*HI C fragment of EBV (see Fig. 3) (YATES et al. 1984). In addition, pBam C type vectors contain the following elements: the β-lactamase gene (*amp*), the pBR*ori* and the Tn5 amino glycoside phosphotransferase gene (*neo*). Thus, mammalian cells carrying pBam C type plasmids can be selected for in G418 containing medium. pBam C is maintained as a plasmid in EBV transformed adherent hybrid human/human cell lines (D98/Raji and D98/HR1). $CaPO_4$ transfection of these cells with pBam C yielded a stable transformation efficiency to G418 resistance of about 0.35%. This transformation efficiency was about 100-fold higher than that in EBV negative cells. Furthermore, the EBV positive cells contained 4–10 episomal copies per cell and no integrated copies of pBam C DNA. When the EBV negative cell lines were transformed to G418 resistance only integrated copies of the *ori* P plasmid were found. Maintenance of pBam C plasmids in EBV positive cell lines requires constant G418 selection; in the absence of selection transformed cells lose the plasmids at 2.5%–6% per generation (see Sect. 3.3). There was no sign of gross rearrangement in plasmids isolated from mammalian cells by the method of HIRT (HIRT 1967) or from *E. coli* which had been transfected with low molecular weight DNA in the HIRT supernatant.

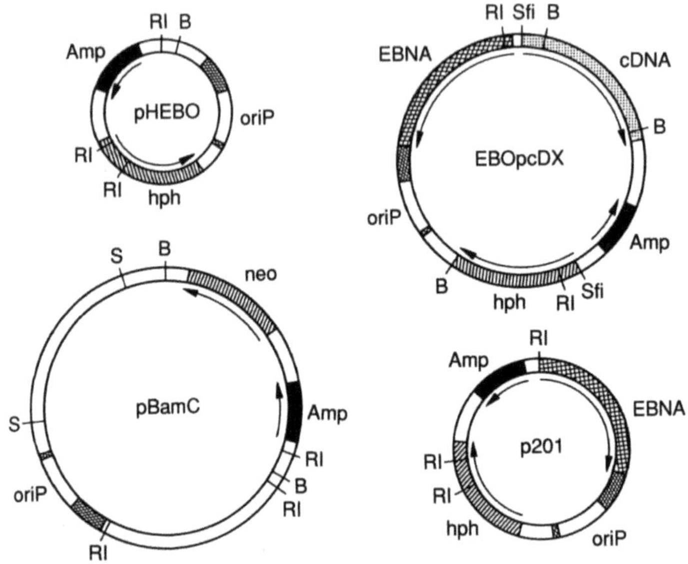

Fig. 3. Maps of EBV based expression vectors. The vector pBam C is 13.8 kb in size and contains the EBV origin for plasmid replication (*ori* P) within the 9.2 kb *Bam*H1 C fragment of EBV (nucleotides 3994 to 13215) and the neomycin (*neo*) gene for resistance to G418 (YATES et al. 1984). The vector pHEBO is 7.2 kb in size and contains *ori* P within a 2.2 kb region of EBV (nucleotides 7315 to 9519) and the hygromycin phosphotransferase (*hph*) gene for resistance to hygromycin B (SUGDEN et al. 1985). The EBV replicon vector p 201 contains *ori* P and *hph* from pHEBO and the EBNA-1 gene (EBV nucleotides 107567–110176) (YATES et al. 1985). The EBV replicon vector EBO-pcD-X contains the same replicon as p 201, but also includes the cDNA cloning and expression elements from plasmid pcD (MARGOLSKEE et al. 1988). The restriction endonuclease sites depicted: *RI*, *Eco*RI; *B*, *Bam*HI; *S*, *Sac* II; *Sfi*, *Sfi* I. The genes depicted: *Amp* (*solid box*), β-lactamase gene of *E. coli*; *ori* P (*fine cross-hatched boxes*), EBV origin for plasmid replication; *hph* (*hatched box*), hygromycin phosphotransferase gene; *neo* (*hatched box*), neomycin phosphotransferase gene; *EBNA* (*cross-hatched box*), EBV nuclear antigen gene; *cDNA* (*stippled box*), pcD transcription unit. *Arrows* indicate direction and extent of transcription. *Vectors* are drawn to scale

A derivative of pBam C, pHEBO (see Fig. 3) (SUGDEN et al. 1985) was developed which contained *ori* P within a 2.2 kb subregion of *Bam*HI C but substituted the hygromycin phosphotransferase (*hph*) gene for *neo*. Transfection of EBV transformed B-lymphoblast cells with pHEBO confers resistance to hygromycin. pHEBO was introduced into B lymphoblastoid cells by electroporation and protoplast fusion. When transfected with pHEBO, 2%–3% of the survivors of the transfection procedure gave rise to hygromycin resistant transformants. The limitation on transfection efficiency was apparently at the stage of DNA entry, since about one-half of those cells which survived transfection and took up DNA became resistant to hygromycin. pHEBO, like pBam C, replicates as a stable plasmid only in EBNA-1 positive cells. Plasmid copy number varied between 1 and 8 for Daudi transformants and 13 and 60 for Raji transformants (SUGDEN et al. 1985). No pHEBO integrants were detected in these transformants. There was a dramatic difference in transformation efficiency in these cells, a factor of more than 10^5, when comparing pHEBO (*ori* P containing) to pHyg (*ori* P lacking). Presumably, this difference relates to the fact that if pHEBO enters the nucleus of an EBNA-1 positive cell it will be maintained as a free plasmid, while pHyg must integrate or be lost. This is a clear example of the enormous power of a replicating plasmid type eukaryotic vector. It is also possible that the high transformation efficiency of pHEBO may in part be due to enhanced transcription of the *hph* due to the EBNA-1 dependent enhancer within *ori* P.

3.2 EBV Replicon Vectors

The *ori* P containing pBam C or pHEBO type plasmids are dependent upon EBNA-1 being supplied in *trans* by the host cell. However, *ori* P plasmids which encode EBNA-1 are independent of host supplied EBNA-1 (YATES et al. 1985; LUPTON and LEVINE 1985). A pHEBO derivative, p201, (see Fig. 3) was constructed wherein the EBNA-1 open reading frame (BKRF-1) was inserted within the pBR322 portion of the plasmid between the β-lactamase gene and *ori* P such that EBNA-1 transcription would be directed toward *ori* P. When p201 (*ori* P/EBNA-1) was transfected into EBNA-1 negative Hela cells a 60-fold increase in hygromycin resistant transformation efficiency (1.8×10^{-3} vs 3.0×15^{-5}) was achieved compared with that with the *ori* P plasmid, pHEBO (YATES et al. 1985).

The EBV DNA containing the EBNA-1 gene in p201 type constructs lacks a promoter: transcription of EBNA-1 in these plasmids is dependent upon the fortuitous expression in mammalian cells of a promoter within the β-lactamase gene of *E. coli*. Plasmid constructs in which the EBNA-1 open reading frame is reversed in orientation compared with p201 do not express EBNA-1 (YATES et al. 1985).

In a different series of constructions (LUPTON and LEVINE 1985), the location of the EBV *ori* P element was confirmed to be within the *Bam*HI C fragment, between nucleotide positions 7315 and 9517. These workers also constructed EBV replicon (*ori* P/EBNA-1) plasmids which demonstrated the necessity of EBNA-1 for episomal maintenance of these plasmids. The basic form of their *ori* P plasmids (p404 and its derivatives) was the insertion of EBV *Bam*HI C subfragments into a pSV2*neo* (SOUTHERN and BERG 1982) derivative between the bacterial β-lactamase gene and the SV40 derived early region. The *ori* P plasmids were maintained episomally in EBNA-1 positive cells at 2–10 copies per cell, and were present only as integrants in EBNA-1 negative cells. Plasmids containing *ori* P yielded 200-fold more G418 resistant transformants in EBNA-1 positive cells than did pSV2*neo* plasmids lacking this element. An EBV replicon plasmid, p410$^+$, was constructed wherein the EBV *Bam*HI K fragment containing the open reading frame for EBNA-1 was inserted downstream of the *neo* gene of p404. The orientation of p410$^+$ is such that EBNA-1 is transcribed in the same direction as *neo*. Although the *Bam*HI K fragment lacks a promoter, EBNA-1 was expressed in this construct. Presumably EBNA-1 mRNA is derived from the same transcription unit as *neo*, either as a dicistronic message or as an alternatively spliced message. The EBV replicon plasmid $p410^+$ is maintained as an episome in both EBNA-1 positive and negative cells. The copy number of this plasmid was between 1 and 10: a 5–10-fold higher copy number occurred in EBNA-1 positive cells than in EBNA-1 negative cells. This effect may relate to low levels of EBNA-1 supplied by this construct.

3.3 Vector Segregation

Both *ori* P and replicon type plasmids are passively lost from cells in the absence of continued drug selection. The rate of loss of pBam C type plasmids (initially present at 4–10 copies per cell) from EBNA-1 positive cells was determined to be about 2.5%–6.0% per generation (YATES et al. 1984; REISMAN et al. 1985). This same rate of loss was found with a replicon type plasmid (initially present at 7–14 copies per cell) in EBNA-1 negative cells (YATES et al. 1985). Plasmids containing one or two functional copies of *ori* P were maintained at a similar copy number and were lost at comparable rates (SUGDEN and WARREN 1988). The fact that double *ori* P plasmids are not unstable suggests that *ori* P does not function like a yeast *cen* element: *cen* leads to plasmid instability when present in multiple copies in a plasmid (HIETER et al. 1985). There is presently no direct evidence pertaining to the rate at which EBV is lost from transformed cells. However, if EBV is more faithful in its segregation than *ori* P plasmids it could be due to a *cen* type element present elsewhere than in *ori* P.

3.4 Vector Encoded Selectable Markers

Several *ori* P and EBV replicon (*ori* P/EBNA-1) plasmids have been developed during the past 4 years. Most of these vectors derive from pBam C or pHEBO (*ori* P type) or p201 (replicon type). Several different genes have been expressed and used as selectable markers for these plasmids. In addition to the *neo* and *hph* genes in the prototype plasmids, the following genes or cDNAs have been cloned in these vectors and selected for by functional expression: hypoxanthine guanine phosphoribosyl transferase (*hprt*) (Y. SHIMIZU et al. 1986; MARGOLSKEE et al. 1988), thymidine kinase (*tk*) (YATES et al. 1985), and Na^+K^+ ATPase (CANFIELD et al. 1990).

3.5 Vector Construction: Orientation and Position Effects

For the most part, EBV vector construction has been done empirically without systematic trials of different configurations. EBNA-1 expression in replicon vectors is dependent upon various heterologous promoters: SV40 late, SV40 early, or a fortuitous promoter within the *E. coli* β-lactamase gene. With all of these vectors the plasmid copy number ranged from 2 to 50. There has not been an attempt to optimize EBNA-1 expression in these vectors. Nor is there any evidence to indicate that the level of vector derived EBNA-1 expression controls plasmid copy number. However, it is noteworthy that a vector containing an EBNA-1 variant lacking most of the alanine-glycine repeat was more effective in promoting *ori* P replication than was wild type EBNA-1 (YATES et al. 1985).

The configuration of vectors with regard to placement and orientation of transcription units, regulatory elements, and replication origins has been arbitrary and empirical. Promoter occulsion, whereby transcription from one promoter preempts transcription from a second nearby promoter (KADESCH and BERG 1986) is a significant consideration in transient expression systems with unintegrated vectors. Upon integration, promoter occlusion is less dramatic. It is uncertain what role, if any, promoter occlusion due to promoter proximity and orientation has in *ori* P or EBV replicon plasmids. Transcription across *ori* P from the EBNA-1 or other transcription units in these vectors may affect vector replication (J.M. YOUNG et al. 1988). However, this has not been systematically tested. Furthermore, elevated levels of replication in SV40 based vectors have been demonstrated to have an adverse effect upon expression of vector encoded genes (LEBKOWSKI et al. 1985). It is uncertain if high levels of *ori* P or *ori* lyt dependent replication would have this same effect.

3.6 Host Range for EBV Vectors

The host range of EBV includes humans and certain other primates. EBV infects only B cells and certain nasopharyngeal cells. However, DNA transfection of EBV replicon plasmids or *ori* P plasmids (in the presence of EBNA-1) has been successful in establishing stably maintained plasmids in several cell lines from a few different species. Monkey kidney cells and dog fibrosarcoma cells are among nonhuman cells which support plasmid replication of EBV vectors. Among human cell lines the following types have been successfully transfected: fibroblastic, epithelial, B-lymphoid (Burkitt's lymphoma and normal B-cells) T-lymphoid, and erythroid cells (YATES et al. 1985; HAMBOR et al. 1988; LUTFALLA et al. 1989). Unfortunately, EBV vectors do not replicate and cannot be maintained as plasmids in any rodent cell lines tested to date (e.g., Balb/C3T3, HTC-A and V79), even when these cells express EBNA-1 (YATES et al. 1985).

It is advantageous to have EBNA-1 expressed from the vector when experiments will utilize several EBNA-1 negative recipients. However, if the recipient is already EBNA-1 positive it is uncertain if any benefit derives from plasmid encoded EBNA-1. Furthermore, depending upon the particular replicon vector and recipient cell, replication may be limited by low levels of EBNA.

The choice of vector and recipient cell obviously depends upon the specific application. Based upon transfection efficiency with *ori* P plasmids the best recipient cells are UC 729-6 (GLASSY et al. 1983; 15%–100% of these cells can be stably transfected by electroporation (MARGOLSKEE et al. 1988; H. HAYAKAWA, personal communication) and an EBNA-1 positive 293 derivative (C.17; 15%–100% of these cells can be stably transfected by $CaPO_4$ (KRYSAN et al. 1989; CANFIELD et al. 1990). Similar transfection efficiencies can be achieved in EBNA-1 negative 293 cells with EBV replicon type vectors. UC 729-6 cells are well suited for purposes which utilize suspension cells (e.g., cell sorting), whereas C.17 or 293 cells are adherent and better suited for applications requiring colony growth. Both of these cell types can generate large numbers of stable transformants (10^6–10^7) from a single transfection.

3.7 *ori* P/*ori* lyt Vectors

Recently, a plasmid, p562, has been developed which contains both the EBV origin for plasmid replication (*ori* P) and the EBV origin for lytic replication (*ori* lyt) (HAMMERSCHMIDT and SUGDEN 1988). This plasmid was contructed by inserting a 5.87 kb segment of EBV DNA containing *ori* lyt into pHEBO. As such, it confers resistance to hygromycin and can be maintained as a low copy plasmid (about 10 copies per cell) in B cell lines which are latently infected with EBV. However, in contrast to *ori* P or *ori* P/EBNA-1 type plasmids, p562 can be induced to undergo the lytic mode of replication. This *ori* P/*ori* lyt plasmid underwent a 400-fold amplification in cells which had been transfected with a pCMV-BZFL-1 plasmid to induce lytic replication. After amplification, most of the p562 DNA was present in the form of high molecular weight concatemers, consistent with its lytic replication via a rolling circle intermediate.

4 EBV Vector Applications

4.1 Analysis of EBV Functional Elements

One of the first applications of EBV vectors was to map and characterize EBV functional elements. Due to the large size of the EBV genome and the lack of a lytic host, elucidation of many EBV functions awaited the cloning and expression of restriction fragments corresponding to the entire EBV genome (SKARE and STROMINGER 1980; DAMBAUGH et al. 1980; BAER et al. 1984; FISCHER et al. 1984; SKARE et al. 1985; TAKADA et al. 1986a; N. SHIMIZU et al. 1988). The seminal work of Sugden and coworkers (YATES et al. 1984) mapped the location of the EBV origin for plasmid replication (*ori* P; see Sect. 2.1). The minimal origin region was identified, and the necessary constituent elements within *ori* P (the tandem repeats and the dyad symmetry element) have been determined. In analogous fashion, the EBV origin for lytic replication (*ori* lyt) has been identified and mapped (HAMMERSCHMIDT and SUGDEN 1988). In addition to origins of DNA replication, EBV enhancers and promoters have also been mapped using *ori* P vectors. EBV replicon vectors were instrumental in demonstrating the *trans*-acting role of the EBNA-1 gene in DNA replication, mapping the required portion of the protein and elucidating the nature of its function in both DNA replication and transcription enhancement (YATES et al. 1985; LUPTON and LEVINE 1985).

4.2 Expression of Foreign Genes and cDNAs

A major application of EBV vectors has been to express foreign genes in mammalian cells. The salient feature of EBV vectors is their ability to be retained as freely replicating plasmids in certain cell lines. This ensures efficient transfection, stable maintenance, and facile recovery of plasmid DNAs. EBV vectors have been used to successfully express several genes and cDNAs. The present list includes: metabolically required enzymes [*tk* (YATES et al. 1985), *hprt* (MARGOLSKEE et al. 1988; Y. SHIMIZU et al. 1986), Na^+K^+ ATPase (CANFIELD et al. 1990)]; drug resistance genes [*neo* (YATES et al. 1984), *hph* (SUGDEN et al. 1985)]; receptors [EGF receptor (YOUNG et al. 1988). FGF receptor (M.C. KIEFER, personal communication), TNF receptor (HELLER et al. 1990)]; cell surface markers [CD8 (MARGOLSKEE et al. 1988), HLA (SHIMIZU et al. 1986), N-CAM (L. PAN and H. BLAU, personal communication)]; peptides [γ-interferon (YOUNG et al. 1988), tumor necrosis factor (KIOUSSIS et al. 1987)]; structural proteins [actin, myosin (L. PAN and H. BLAU, personal communication)]; and viral genes [V-*myc* (LOMBARDI et al. 1987), HIV-*tat* (FENG and HOLLAND 1988), influenza HA (JALANKO et al. 1989)]. In those cases where it was determined, plasmids were stably maintained as unrearranged episomes, with 2–50 copies per cell. Constructs contained 1–3 mammalian transcription units and inserts up to 35 kb in length. Taking into consideration the fact that EBV itself is a 172 kb plasmid, it is possible that extremely large inserts may be accommodated in EBV based plasmids. This situation is in marked contrast to BPV based vectors, wherein several different inserts lead to deletion, rearrangement, recombination, or integration (for examples see MITRANI-ROSENBAUM et al. 1983; WANG et al. 1983; ZINN et al. 1983; DIMAIO et al. 1984; HSIUNG et al. 1984; LEHN and SAUER 1984; PINTEL et al. 1984).

EBV *ori* P vectors encoding class I HLA (A2 and B8) genes were used to express these genes in mutant B-lymphoblastoid cells deficient in HLA expression (Y. SHIMIZU et al. 1986). Cell surface HLA expression was followed by flow cytometry. In this way, mutations were classified as being in either structural or *trans*-acting genes required for HLA expression. In this system, the copy number of the plasmid was responsive to the hygromycin concentration in the medium: 1–2 copies per cell at low drug concentrations, 5–10 copies per cell at high concentrations. However, cell surface expression of HLA was independent of copy number, suggesting that some factor other than copy number is limiting expression. In this regard, other investigators have found that high levels of hygromycin in the medium may lead to deletion of inserts from EBV replicon vectors (P. KAVATHAS and R.F. MARGOLSKEE, unpublished work).

Another cell surface marker which has been expressed from *ori* P and replicon vectors is CD8 (Leu-2; MARGOLSKEE et al. 1988; HAMBOR et al. 1988). These plasmids use the SV40 or RSV promoter to express a CD8 cDNA. Different cDNA isolates express either a secreted or membrane bound form of CD8 (P. KAVATHAS and R.F. MARGOLSKEE, unpublished work). In a recent study, a chimera was constructed which linked the extracellular domain of CD8 with the glycolipid anchoring structure (PIG tail) from decay-accelerating factor (HAMBOR et al. 1988). In this system, CD8 is used as a "reporter" epitope to identify and analyze the anchoring signals. Another study used the CD8 cDNA in antisense orientation in an EBV replicon plasmid to follow the effects of antisense RNA mediated gene inhibition of T cell functions (TYKOCINSKI et al. 1988). Antisense RNA expressed from an EBV replicon plasmid was >95% effective in inhibiting cell-surface CD8 expression. The effect was highly selective. Reversal of the antisense inhibition was achieved by removing the transfected cells from selective medium. Upon passive loss of the plasmid, CD8 returned to the cell surface.

A systematic survey of promoter activity in transcribing the CAT gene was undertaken in human T lymphocytes (HAMBOR et al. 1988). Promoter-CAT cartridges were inserted into a p201 derivative replicon plasmid between the 3′ end of the *hph* gene and the tandem repeat element of *ori* P. Both cartridge orientations were assayed for CAT activity. In this series of constructs the rank order of promoters in producing CAT activity was: RSV ≧ lymphopapilloma virus > metallothionein (plus Cd^{++}) > SV40 early > metallothionein (minus Cd^{++}).

An EBV replicon type plasmid was constructed to assay γ-interferon (γ-IFN) expression from various promoters (J.M. YOUNG et al. 1988). This plasmid had a configuration similar to that of the replicon CAT plasmids: the γ-IFN transcription unit was interposed between *ori* P and the *hph* gene. Three different promoters were evaluated: an enhancer-less metallothionein promoter (Pmt), the metallothionein promoter coupled to the MSV enhancer (Emsv Pmt), and the SV40 promoter linked to the CMV enhancer (Ecmv Psv). When these constructs were assayed in CV-1 cells the following rank order of promoter activity was found: Ecmv Psv ≫ Emsv Pmt ≈ Pmt. The most striking finding was that all three types of promoter-γ-IFN cartridge inserts were only active in that orientation in which their transcription was directed away from *ori* P and toward the convergent transcription from the *hph* gene (i.e., EBNA → *ori* P IFN → ← *hph* was active whereas EBNA → *ori* P ← IFN ← *hph* was inactive). Preliminary results suggested that the orientation effect decreased γ-IFN expression via decreased RNA levels. It was not determined if this was due to decreased transcription, unstable mRNA, or aberrant processing. Several different mechanisms might explain the orientation effect. It is possible that γ-IFN transcription across *ori* P inhibits replication of these plasmids.

Whereas the copy number of the expressing plasmids was 1–5 episomes per cell, the copy number and the state of DNA for the reverse orientation (nonexpressing) plasmids was not determined. An alternative explanation for the effect of cartridge orientation on expression could be interference with EBNA-1 transcription since these transcription units converge upon *ori* P. However, this seems unlikely since the other orientation presumably would have interfered with *hph* expression and this was not a problem. In addition, the orientation effect could be due to the juxtaposition of the γ-IFN transcription unit to the *ori* P tandem repeats (an EBNA-1 dependent enhancer) in the positive case, whereas the reverse (unexpressed) orientation uncouples the γ-IFN transcription unit from the *ori* P enhancer. Finally, it is possible that promoter occlusion (see KADESCH and BERG 1986) whereby an upstream promoter reduces enhancer dependent transcription of a downstream promoter, or "promoter damping" whereby an adjacent or intervening promoter decreases enhancer transcription could be reponsible for this orientation effect. Promoter occlusion and promoter damping have only been studied with linearly arrayed promoters. These effects have not been studied in EBV replicon plasmids, nor when transcription units are arrayed in convergent or divergent orientations (as with most EBV replicon plasmids). Whatever the explanation, it seems likely that some unique feature of the γ-IFN cartridge itself interferes with expression or replication when it is in one orientation. Several other examples of analogous constructs with many different combinations of promoters and reporter genes inserted at the same site of EBV replicon plasmids do not show this pronounced orientation effect (HAMBOR et al. 1988; MARGOLSKEE et al. 1988).

4.3 EBV Based Shuttle Vectors to Study Mutagenesis

Originally, SV40 based shuttle vectors were developed which allowed the introduction of specific bacterial genes as targets for mutagenesis in mammalian cells. SV40 vectors undergo transient replication in permissive cells, and can be analyzed for mutations after recovery by "shuttling" back into *E. coli*. However, SV40 vectors acquire mutations in their target genes at the level of 1% simply by the process of mammalian cell transfection (CALOS et al. 1983; RAZZAQUE et al. 1983). The following unique features of *ori* P and EBV replicon vectors make them ideally suited for the study of mutagenesis in human cells. (a) Stable plasmid replication of EBV vectors allows long-term experiments with clonal cell lines followed by the ready recovery of mutagenized DNAs. (b) EBV vectors have very low levels of spontaneously induced mutations. (c) These vectors appear to replicate only once per cell cycle in synchrony with host DNA. In contrast, shuttle vectors based upon BPV replicons have been characterized by high inherent mutation frequencies, even in clonal cell lines (ASHMAN and DAVIDSON 1985). Mutations were mainly due to rearrangements and presumably reflect the underlying instability of BPV based vectors.

An *ori* P vector containing the HSV *tk* gene (DRINKWATER and KLEINDINST 1986) and an EBV replicon vector cotaining the *E. coli lac* I gene (DUBRIDGE et al. 1987) have been used as shuttle vectors to assay mutagenesis in human cells. As with the SV40 based transient vectors, EBV shuttle vectors undergo a significant level of mutagenesis upon transfection: from 3×10^{-5} to 6×10^{-5} for the *tk* gene and 1×10^{-2} to 5×10^{-4} for the *lac* I gene. Mutation frequency depended in large part upon the host cell. Although, the mutation frequency of EBV shuttle vectors after transfection was not significantly less than that of the SV40 based vectors, it was possible to isolate stable clonal transformants in which the EBV shuttle vector had not been mutagenized. This cannot be done with an SV40 based system. Clonal isolates with mutation frequencies of 6×10^{-6} per 1000 bp of the *lac* I genes, or 6×10^{-9} per base pair were found. This is similar to the frequency of spontaneous mutations in human and bacterial genes. Chemical mutagenesis of clonal cells containing an EBV shuttle vector demonstrated a 300-fold increase in mutation frequency in response to N-nitroso-N-methylurea (NMU). Furthermore, recovery of EBV shuttle vectors from NMU mutagenized cells demonstrated that all 270 *lac* I mutations analyzed had base substitutions, with a predilection for G·C to A·T transitions, particularly within regions of consecutive G·C base pairs.

A modification of the EBV shuttle system utilizing T antigen dependent SV40 *ori* replication has been employed to increase vector copy number (HAASE et al. 1989; HEINZEL et al. 1988). EBV vectors typically have a copy number between 1 and 100, whereas SV40 *ori* plasmids can achieve 10000–100000 DNA molecules per cell in permissive cells. When a clonal line of human cells containing an EBV shuttle vector with an SV40 *ori* was transfected with a nonreplicating plasmid encoding SV40 T antigen, an increase in copy number to about 1000 EBV plasmids per cell was achieved. This boost only occurred in those cells which had received T antigen, and since the T antigen was supplied in *trans* from a plasmid which was passively lost, the increased copy number returned toward initial preboost levels over a period of several days. It is at present unclear why the boosted copy number cannot be maintained. Perhaps there is a selective disadvantage for such high numbers of EBV replicon plasmids or the EBV *ori* P/EBNA-1 dependent replication machinery cannot accommodate so many copies per cell. The SV40 boost did not raise the point mutation frequency of the *lac* I gene. However, deletions and in certain cases intramolecular recombination increased signifcantly after the boost.

If the SV40 boost raises the expression of genes encoded within the EBV replicon vectors, then this method may have utility for gene expression and expression cloning. In at least one case, SV40 replication was found to be inhibitory to expression of genes encoded within an SV40 vector (LEBKOWSKI et al. 1985). However, viral encoded T antigen boost of CV-1 cells stably transfected with the SV40 expression plasmid pCD led to excisional replication of the plasmid to high copy number and a concomitant 20-fold increase in expression of the pcD encoded gene (M MANOS, personal communication; WEAVER et al. 1988). The effect of T antigen induced amplification of SV40/EBV hybrid replicons on gene expression has not been specifically examined.

4.4 EBV Based Expression Cloning Vectors

There are presently only a few general strategies for cloning a eukaryotic gene. Only rarely is the mRNA of the gene so highly expressed in a tissue or cell that it may be physically purified and converted into a relatively pure cDNA clone. However, the vast majority of mammalian mRNAs are of such low abundance that this direct approach cannot usually be applied. For a few low-abundance mRNAs, conditions for differential expression allow the generation of a cDNA probe enriched for the sought-after gene. In situations where neither RNA nor enriched cDNA is available but the protein product of the gene is abundant, the following approach has often been successful: the protein is purified, a partial amino acid sequence is determined, oligonucleotide probe(s) based upon the peptide sequence are synthesized and then used to screen an appropriate library for a clone that hybridizes with the probe(s).

Within the past few years various cloning strategies based upon expression have been utilized to clone several genes. Expression cloning employs specialized vectors designed for the particular recipient cell and the different types of expression cloning can be distinguished by the recipient cell used: *E. coli* or mammalian cells. For expression cloning in *E. coli*, λgt11 has been used to express cDNA clones as β-galactosidase fusion products (RA Young and Davis 1983). λgt11 clones are screened with antibodies directed against the gene of interest. Expression cloning in mammalian cells has been accomplished by the use of the pcD vectors (Okayama and Berg 1982, 1983, 1985). The pcD plasmid is an SV40 based integrating-type mammalian expression vector: the vector supplies all the needed elements for proper expression in mammalian cells.

Expression cloning from pcD libraries does not rely upon the availability of purified protein, protein or nucleic acid sequence data, or specific antibodies. Rather, pcD expression cloning depends upon the ability to select or screen for the functional expression of the gene in mammalian cells. The limitations of this approach are the requirement for full-length (or nearly full-length) cDNA clones for expression, the necessity of obtaining sufficient mammalian cell transformants to provide a full representation of the cDNA library, and the need to recover the integrated cDNA sequence from the mammalian genome. The latter two difficulties have been alleviated by the use of an EBV replicon vector, EBO (see Fig. 3) for cDNA expression libraries (Margolskee et al. 1988). The EBO vector contains *ori* P, EBNA-1 and *hph*. This vector allows the high efficiency transformation of human cells with an entire pcD cDNA library. Furthermore, as is the case with other EBV replicon vectors, the EBO-pcD library replicates as autonomous plasmids in EBV-transformed recipient cells. EBO-pcD has been used to isolate clones for the cell surface marker CD8 and for the enzyme HPRT. The ability to directly select for expression of very rare EBO-pcD clones and to recover episomes facilitated the cloning of these genes.

Efficient transfection of the EBO-pcD-vector into human lymphoblastoid (UC 729-6) cells was achieved by electroporation (Potter et al. 1984). Using the plasmid EBO-pcD-CD8 (an EBO-pcD clone expressing CD8), about 20% of the cells survived the electroporation and about 33% of the survivors expressed

CD8 on the cell surface after 48 h (MARGOLSKEE et al. 1988). After selection in hygromycin-B-containing medium, a pure population of hygromycin-resistant cells expressing CD8 was obtained. Of those cells that had survived the electroporation, 15% were transformed to hygromycin resistance: (i.e., from 5×10^7 UC cells transfected with EBO-pcD-CD8, about 1.5×10^6 hygromycin-resistant, CD8-expressing cells were obtained). These cells were shown to carry eight to ten copies of the EBO-pcD-CD8 plasmid, with no detectable integrations or rearrangements (MARGOLSKEE et al. 1988).

The sensitivity of the EBO-pcD vector for cloning rare cDNA clones was illustrated by several reconstruction experiments. It was possible to clone EBO-pcD-CD8 by expression from cells which had received a 1:20000 mixture of this plasmid with EBO-pSV2*neo* (a homologous construct which express *neo*; MARGOLSKEE et al. 1988). Four sequential cycles of cell sorting by a fluorescence activated cell sorter (FACS) for the CD8 positive fraction of stable transformants from this transfection yielded a pure population of CD8 positive cells. The sorted cells contained approximately ten copies per cell of EBO-pcD CD8, two of which comigrated with monomer plasmids and eight of which comigrated with high-molecular-weight DNA (MARGOLSKEE et al. 1988). The high-molecular-weight form of EBO-pcD CD8 was thought to be due to concatemerization rather than integration. This apparent concatemerization is the result of the FACS selection process and not an inherent property of the EBO plasmid: UC cells transfected with EBO-pcD-CD8 but not subjected to FACS maintain the plasmid as an unintegrated monomer. The low-molecular-weight plasmid DNA from these CD8 expressing cells was used to transform *E. coli*. Five of six plasmid DNAs recovered in this way were identical to the input plasmid and were capable of transforming cells to express CD8. EBO-pcD-CD8 was also cloned directly from an EBO-pcD library (MARGOLSKEE et al. 1988). These results demonstrate that EBO plasmids can be recovered from libraries or artificial mixtures—a necessary prerequisite for eukaryotic expression cloning.

EBO-pcD expression was also used to clone a cDNA for the metabolically required enzyme HPRT (MARGOLSKEE et al. 1988). Transfection of 5×10^7 HPRT negative UC cells with EBO-pcD*hprt* yielded a hygromycin resistance transformation frequency of 10%–15% and a HAT resistance transformation frequency of 5%–10%. DNA mixtures at ratios from 1:20000 to $1{:}10^6$ of EBO-pcD*hprt* and EBO-pSV2*neo* were introduced by electroporation into UC cells. Survivors of the electroporation were selected in hygromycin-B or HAT-containing media. HPRT-positive transformants were obtained irrespective of whether cells were selected initially in hygromycin-B medium or directly in HAT medium (Tables 1 and 2). Plasmid DNA and total genomic DNA were isolated from UC cells that had received the 1:200000 mixture and had been subjected to either the two-step selection (hygromycin followed by HAT) or the one-step selection (HAT alone). In these transformants, EBO-pcD*hprt* was present as an unrearranged extrachromosomal plasmid at between two and eight copies per cell. During 30–50 generations in HAT-selective media, the representation of EBO-pcD*hprt* DNA in the HPRT positive transformants increased relative to

Table 1. Direct HAT-resistance selection of UC lymphoblastoid cells transfected with EBO-pcD*hprt* and EBO-pSV2*neo* DNAs mixed at various ratios

Plasmid ratio (EBO-pcD*hprt*/EBO-pSV2*neo*)[a]	HAT-esistant colonies/5×10^6 cells[b]	HAT-resistant colonies/2×10^7 cells[b]	Predicted HAT-resistance frequency[c]
0: 20000	0	0	0
1: 20000	+	+	3.5×10^{-6}
1: 200000	+	+	3.5×10^{-7}
1:1000000	+	+	7.0×10^{-8}

[a] The total amount of mixed plasmid DNA used for transfection was 100 μg per 5×10^7 UC cells
[b] 5×10^6 or 2×10^7 UC cells which survived electroporation were selected en masse in flasks containing HAT medium
[c] The expected HAT-resistance frequency was calculated based upon the known hygromycin resistance frequency (7.0%) and the ratio of EBO-pcD*hpt* to EBO-pSV2*neo* in the transfected DNA

Table 2. HAT-resistance selection following pre-selection in hygromycin B of UC lymphoblastoid cells transfected with EBO-pcD*hprt* and EBO-pSV2*neo* DNAs mixed at various ratios

Plasmid ratio (EBO-pcD*hprt*/EBO-pSV2*neo*)[a]	HAT-resistance frequency[b]	Predicted HAT-resistance frequency[c]
0: 20000	0	0
1: 20000	2×10^{-5}	5×10^{-5}
1: 200000	3×10^{-6}	5×10^{-6}
1:1000000	5×10^{-7}	1×10^{-6}

[a] The total amount of mixed plasmid DNA used for transfection was 100 μg per 5×10^7 UC cells
[b] Determined by dilutional plating of hygromycin resistant transformants into HAT medium
[c] Calculated based upon the ratio of EBO-pcD*hprt* to EBO-pSV2*neo* in the transfected DNA, assuming entry of one EBO plasmid per cell

the EBO-pSV2*neo* DNA by a factor of 10^6–10^7 (Margolskee et al. 1988). Plasmid DNA isolated from these HAT-resistant transformants was used to transform *E. coli* DH5. (Hanahan 1983). All of the rescued colonies contained plasmid DNAs with a restriction pattern characteristic of EBO-pcD*hprt*. Reintroduction of the recovered DNAs into UC cells yielded HAT-resistant transformants. These experiments demonstrated that the EBO-pcD expression vector can be used to directly select and recover very rare clones (10^{-6}) that code for a metabolically required enzyme.

The EBO-pcD vector improves the pcD expression cloning system to allow for rapid selection, cloning, and recovery of cDNA clones. A schematic representation of how EBO-pcD-cDNA expression cloning could be utilized to clone cell-surface markers and receptors is shown in Fig. 4. The salient features of this approach are as follows (a) an EBO-pcD library of several million

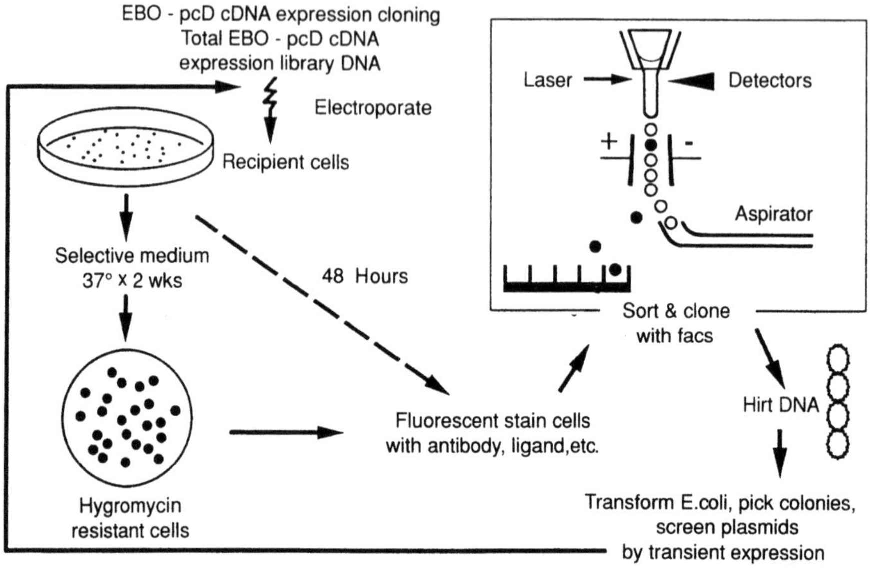

Fig. 4. EBO-pcD-cDNA expression cloning of cell surface receptors (see text for details)

independent bacterial clones in generated using mRNA from a cell line expressing the receptor gene to be cloned; (b) DNA from the total EBO-pcD-cDNA expression library is introduced by electroporation into a receptor-negative recipient cell line; (c) several million independent mammalian cell transformants are selected by growth in hygromycin-B-containing medium; (d) hygromycin-resistant transformants (or unselected cells at 48 h after transfection) are screened for expression of cell-surface receptors by the binding of fluorescent ligand or fluorescent antibody, followed by sorting with the FACS; (e) repeated cycles of sorting with the FACS are used to isolate those rare transformants expressing cell-surface receptor; (f) EBO-pcD plasmids are rescued from receptor-positive cells by isolation of low-molecular-weight DNA which is used to transform competent *E. coli*; (9) EBO-pcD plasmids recovered from individual bacterial colonies are screened for the induction of expression of receptors by reintroduction of the DNAs into mammalian cells. Utilizing a fluorescent antibody, this cloning procedure was used to clone CD8 (MARGOLSKEE et al. 1988). Recently, employing a fluorescent ligand (fluorescein-TNF) this same procedure has been used to clone the TNF receptor (HELLER et al. 1990; M. GARRET, B. RINDE, N. SPICKOFSKY, R.F. MARGOLSKEE, in preparation).

4.5 EBV Based Cosmid Cloning Vectors

Several eukaryotic genes have been cloned by expression using the following approach. (a) Mammalian cells are transfected with randomly sheared high-molecular-weight genomic DNA. (b) Transformants are selected by cotransfection with an integrating vector encoding drug resistance (e.g., pSV2*neo*). (c) Transformants are screened for expression by staining with antibodies directed against the gene to be cloned. (d) Positively expressing transformants are selected by multiple rounds (usually three to five cycles) of enrichment via FACS, antibody rosetting or antibody panning. (e) Genomic DNA is isolated from positive transformants and reintroduced into nonexpressing cells. (f) Steps (b) through (e) are repeated as necessary to limit the amount of foreign DNA to a manageable amount. (9) Genomic libraries are constructed from the positive transformants and screened for repetitive sequence elements from the species of origin for the heterologous DNA (e.g., *Alu* sequences). This paradigm has often been successful, but it is quite time consuming and tedious (for examples see: KAVATHAS et al. 1984; LITTMAN et al. 1985).

One alternative to the genomic transfection/screening approach is the EBO-pcD expression cloning system described in Sect. 4.4 of this review. Another alternative has recently been developed which utilizes genomic libraries prepared in a cosmid derivative of an EBV replicon vector (KIOUSSIS et al. 1987). These authors inserted λphage cohesive ends (cos) into the EBV replicon vector p201 to generate cos 202. Human DNA partially digested with *Sau* 3A (35–40 kb size fraction) was inserted into cos 202 via its unique *Bgl* II site. A cosmid library of 10^6 independent clones was generated, inserts averaged about 35 kb (this yields a library representing about ten fold the human haploid genome). Human 143/EBNA$^+$ cells were transfected with cosmid library DNA and selected for hygromycin resistance to yield 40000 transformants. Transformants had 1–10 copies of cosmid DNAs per cell, this would be equivalent to 0.5- to 5-fold the human haploid genome, if the 10 copies per cell are due to different plasmids. Supercoiled plasmid DNA was isolated from the transformants containing the library, and used to successfully transfect *E. coli* to ampicillin resistance (recovering 100–1000 colonies with DNA from 10^7 human transformants). Cosmid inserts were 25–35 kb in size for the *E. coli* rescued clones. In sum, it was possible to generate an EBV based cosmid library, introduce it into human cells, and recover clones in *E. coli*.

In a further test, EBV cosmids encoding the tumor necrosis factor (TNF) gene were identified by oligonucleotide hybridization. TNF specific EBV cosmids were introduced into 143/EBNA$^+$ cells where TNF specific mRNAs were generated (KIOUSSIS et al. 1987). S1 analysis indicated proper 5′ and 3′ ends for the TNF mRNA. Preliminary data suggested that TNF protein was present and biologically active. There were 2–50 copies of TNF specific mRNA in human 143/EBNA$^+$ cells transformed by the TNF cosmid (untransfected controls did not express this RNA). The TNF cosmids could be rescued from 143/EBNA$^+$ cells by isolating supercoiled DNA from their nuclei, packaging it into phage particles, and infecting *E. coli*. About 90% of the rescued cosmids were identical to the input cosmid transfected into 143/EBNA$^+$ cells. However, a few cosmids were recovered that had rearranged with a loss of repetitive DNA. Presumably this occurred during replication in the human cell line.

In this study it was necessary to use an EBNA-1 positive cell line as a recipient to achieve maximum transformation efficiency (Kioussis et al. 1987). When EBNA-1 negative HeLa cells were transfected with cos 202 lacking an insert, $\sim 10^3$ colonies/5 μg DNA/5×10^5 cells were obtained. However, when cos 202 contained an insert only 10 colonies/5 μg DNA/5×10^5 cells were obtained. This decreased efficiency could be overcome by using EBNA-1 positive HeLa cells: transfection of these cells with cos 202 containing an insert yielded about 10^3 colonies/5 μg DNA/5×10^5 cells. Presumably, inserting 35 kb of genomic DNA in cos 202 decreased the level of EBNA-1 expression from this cosmid. It might be possible to rectify this by placing a very strong promoter in front of the EBNA-1 gene. These authors were not able to rule out integration events; however, at least some of the cosmids were maintained as plasmids (Kioussis et al. 1987).

Although in principle this cosmid should provide a means to clone genes from a library by expression, this study did not directly demonstrate that this can be done (as has been done for the EBO-pcD vectors). It is possible that expression from the cosmid vectors may be inadequate to identify clones using fluorescent tagged antibodies. Furthermore, it may not be possible to recover one cosmid clone from the 10^6 in the original library if preliminary stages in expression lead to rearrangement or integration. Transfection and maintenance of entire EBV cosmid libraries may be difficult. Prior to using this method for gene cloning by expression, a test case similar to that used for the EBO-pcD system should be done. Assuming the EBV cosmid system does work for expression cloning, then it may provide a viable alternative to the EBO-pcD method in some cases. One weakness of cDNA expression cloning methods is the requirement for full length expressible cDNAs. This becomes problematic when the mRNA is extremely long, quite rare, or displays secondary structures inhibitory to successful full-length reverse transcription. The cDNA cloning approaches are likely to be most successful for abundant, short mRNAS. EBV based cosmid expression cloning with cos 202 may be successful even if the mRNA is long and rare. However, if the gene is extremely large (> 35 kb) and if it has a very weak promoter, cos 202 expression cloning may not succeed. Incorporation of a strong promoter and/or enhancer in the EBV cos 202 vector upstream or downstream of the cloned DNA dramatically improves expression of genomic inserts (S. McLlaughlin and R.F. Margolskee, in preparation).

5 Future Directions

Presently, much work is being devoted to mapping and analyzing the interaction of *cis*-acting *ori* lyt and *trans*-acting ZEBRA protein. The practical value of these studies is the eventual incorporation of an inducible lytic replicon into the next generation of EBV vectors. This would allow the controlled amplification of

EBV replicons, which in turn might enhance expression (via increased copy number), aid recovery of cloned genes, and make shuttle vectors for the analysis of mutagenesis more sensitive. However, for these vectors to be useful the boost must not interfere with expression. It is unclear at present what effect rolling circle type replication of *ori* lyt containing vectors will have upon expression of vector encoded genes. Lytic replication of the resident EBV and the formation of recombinants with the EBV plasmid may present problems. Furthermore, packaging of mature EBV virus or recombinants may occur.

SV40 *ori* induction should be considered as an alternative to amplification by the *ori* lyt/ZEBRA replicon. The SV40 *ori*/T antigen replicon is capable of replicative amplification even greater than that of *ori* lyt, and the SV40 replicon is already the best characterized of any eukaryotic virus (Shortle and Nathans 1979; Shortle et al. 1979). T antigen can be readily supplied by superinfection with adenovirus/SV40 hybrids, wild type SV40, or *ori* defective SV40 (Gluzman 1981; Margolskee and Nathans 1984). Alternatively, transfection with plasmids encoding T antigen can be used to efficiently boost SV40 *ori* based replicaton. This approach has been successfully used for EBV shuttle vectors (Haase et al. 1989; Heinzel et al. 1988).

Most of the EBV vectors which have been used were constructed based upon empirical considerations. Few systematic studies of EBV vector structure have been undertaken. Possible sources for improvement are the following. (a) Strong promoters (e.g., CMV or RSV) adjacent to cDNA inserts may yield better expression and expression cloning vectors. (b) Orientation and location of inserts must be so as to not interfere with EBNA-1 or *ori* P functions. (c) Cosmid vectors would be much improved by the addition of strong enhancers flanking the genomic DNA insert.

One application for which EBV vectors are ideally suited is the analysis of expression of mutagenized inserts. Both directed mutagenesis and random mutagenesis could be used to analyze genes or cDNAs in EBV vectors. The following directed mutagenesis approach should be very productive. (a) Oligonucleotide mutagenesis of the target gene is carried out in M13. (b) The mutagenized target gene is transferred to an EBV vector. (c) The collection of mutagenized genes in the EBV vector is transfected and analyzed for altered expression. This approach has recently been used to identify the specific changes required to convert a ouabain sensitive isoform of the Na^+K^+ ATPase to ouabain resistance (Canfield et al. 1990). This approach could be utilized for any target gene where altered expression can be assayed. Alternatively, random mutagenesis can be employed, wherein a collection of discrete, localized, random mutations are introduced within the target gene (Shortle and Nathans 1978; Shortle et al. 1979). Successful random approaches are dependent upon a biological selection to identify and enrich the mutants (e.g., ouabain resistance of the Na^+K^+ ATPase). The directed and random mutagenesis approaches can be used in concert to yield complementary information.

Expression cloning is a relatively new application for EBV replicon vectors. Both genomic and cDNA type expression vectors have been developed. It is likely that this will become a major use of these vectors. Potential targets for expression cloning utilizing EBV vectors are: cell surface markers, receptors, ion channels, ion pumps, and *trans*-acting regulatory factors. These genes or cDNAs would also be good targets for EBV replicon mutagenesis.

Although EBV vectors have a wide host range they do not function as replicons in any rodent cells tested to date. Considering the number of well-characterized mutant cell lines derived from rodents, it would be extremely useful to have a functional episomal expression vector system for these cell lines. Unfortunately, although BPV derived vectors have a host range which includes rodent cells, these vectors are not reliably maintained as episomes, and often undergo rearrangements. It might be possible to produce human/hamster hybrid cells which would accommodate stable plasmid maintenance of EBV replicons. Furthermore, it may be possible to identify and clone the human permissive factors needed for EBV episome replication which are lacking in rodent cell lines.

6 Conclusions

EBV vectors are still in an early stage of development. However, the stable plasmid maintenance of these vectors fills an important niche for many vector applications.

Important applications include eukaryotic expression, shuttle mutagenesis, expression cloning, and directed and random mutagenesis. As more experience is gained with these vectors, their sophistication will grow and it will be possible to more accurately predict their behavior. The unique nature of EBV plasmid vectors will lead to greatly expanded use of those vectors. The strengths inherent to EBV plasmid vectors are the following. (a) These vectors can be readily introduced into mammalian cells with very high efficiency. (b) EBV vectors are maintained stably as plasmids without deletion, integration, or rearrangement. (c) These vectors can be easily recovered from transfected cells by shuttling into *E. coli*.

References

Adams A (1987) Replication of latent Epstein-Barr virus genomes in Raji cells. J Virol 61: 1743–1746

Adams A, Lindahl T (1975) Epstein-Barr virus genomes with properties of circular DNA molecules in carrier cells. Proc Natl Acad Sci USA 72: 1477–1481

Ashman CR, Davidson RL (1985) High spontaneous mutation frequency of BPV shuttle vector. Somate Cell Mol Genet 11: 499–504

Baer B, Bankier A, Biggin M, Deininger PL, Farrell PJ, Gibson TJ, Hatfull G, Hudson GS, Satchwell SC, Seguin C, Tuffnell PS, Barrell BG (1984) DNA sequence and expression of the B95-8 Epstein-Barr virus genome. Nature 310: 207–211

Baichwal VR, Sugden B (1988) Latency comes of age for herpesviruses. Cell 52: 787–789

Birx DL, Redfield RR, Tosato G (1986) Defective regulation of Epstein-Barr virus infection in patients with acquired immunodeficiency syndrome (AIDS) or AIDS-related disorders. N Engl J Med 314: 874–879

Bodescot M, Perricaudet M (1986) Epstein-Barr virus mRNA produced by alternative splicing. Nucleic Acids Res 14: 7103–7114

Bodescot M, Brison O, Pericaudet M (1986) Epsten-Barr virus transcription unit is at least 84 kilobases long. Nucleic Acids Res 14: 2611–2620

Bodescot M, Perricaudet M, Farrell PJ (1987) A promoter for the highly spliced EBNA family of RNAs of Epstein-Barr virus. J Virol 61: 3424–3430

Calos MP, Lebkowski JS, Botchan MR (1983) High mutation frequency in DNA transfected into mammalian cells. Proc Natl Acad Sci USA 80: 3015–3019

Canfield V, Emanuel JR, Spickofsky N, Levenson R, Margolskee RF (1990) Ouabain resistant mutants of the rat Na, K-ATPase α2 isoform identified using an episomal expression vector. Mol Cell Biol 10: 1367–1372

Chevallier-Greco A, Manet E, Chavrier P, Mosnier C, Daillie J, Sergeant A (1986) Both Epstein-Barr virus (EBV)-encoded *trans*-acting factors, EB1 and EB2, are required to activate transcription from an EBV early promoter. EMBO J 5: 3243–3249

Chittenden T, Lupton S, Levine A (1989) Functional limits of *ori* P, the Epstein-Barr virus plasmid origin of replication. J Virol 63: 3016–3025

Cleary M, Dorfman RF, Sklar J (1986) Failure in immunological control of the virus: post-transplant lymphoma. In: Epstein MA, Achong BG (eds) The Epstein-Barr virus: recent advances. Wiley, New York

Countryman J, Jenson H, Seibl R, Wolfe H, Miller G (1987) Polymorphic proteins encoded within BZLF1 of defective and standard Epstein-Barr viruses distrupt latency. J Virol 61: 3672–3679

Dambaugh T, Biesel C, Hummel M (1980) Epstein-Barr virus (B95-8) DNA VII: molecular cloning and detailed mapping. Proc Natl Acad Sci USA 77: 2999–3003

Dambaugh T, Hennessy K, Chamnankit L, Kieff E (1984) U2 region of Epstein-Barr virus DNA may encode Epstein-Barr nuclear antigen 2. Proc Natl Acad Sci USA 81: 7632–7636

Dillner J, Kallin B, Alexander H, Ernberg I, Uno M, Ono Y, Klein G, Lerner RA (1986) An Epstein-Barr virus (EBV)-determined nuclear antigen (EBNA5) partly encoded by the transformation-associated Bam WYH region of EBV DNA: preferential expression in lymphoblastoid cell lines. Proc Natl Acad Sci USA 83: 6641–6645

DiMaio D, Corbin V, Sibley E, Maniatis T (1984) High-level expression of a cloned HLA heavy chain gene introduced into mouse cells on a bovine papillomavirus vector. Mol Cell Biol 4: 340–350

Drinkwater NR, Kleindinst DK (1986) Chemically induced mutagenesis in a shuttle vector with a low-background mutant frequency. Proc Natl Acad Sci USA 83: 3402–3406

DuBridge RB, Tang P, Hsia HC, Leong P-M, Miller JH, Calos MP (1987) Analysis of mutation in human cells by using an Epstein-Barr virus shuttle system. Mol Cell Biol 7: 379–387

Epstein MA, Achong BG (1979) The relationship of the virus to Burkitt lymphoma. In: Epstein MA, Achong BG (eds) The Epstein-Barr virus. Springer, Berlin Heidelberg New York, pp 321–327

Feng S, Holland EC (1988) HIV-1 *tat trans*-activation requires the loop sequence within *tar*. Nature 334: 165–167

Fingeroth JD, Weis JJ, Tedder TF, Strominger JL, Biro PA, Feron DT (1984) Epstein-Barr virus receptor of human B lymphocytes is the C3d receptor CR2. Proc Natl Acad Sci USA 81: 4510–4514

Fischer DK, Robert MF, Shedd D, Summers WP, Robinson JE, Wolak J, Stefano JE, Miller G (1984) Identification of Epstein-Barr nuclear antigen polypeptide in mouse and monkey cells after gene transfer with a cloned 2.9-kilobase-pair subfragment of the genome. Proc Natl Acad Sci USA 81: 43–47

Frade R, Barel M, Ehlin-Henriksson B, Klein G (1985) gp140, The C3d receptor of human lymphocytes, is also the Epstein-Barr virus receptor. Proc Natl Acad Sci USA 82: 1490–1493

Fresen KO, Merkt G, Bornkamm GW, zur Hausen H (1977) Heterogeneity of Epstein-Barr virus originating from P3HR-1 cells. I. Studies on EBNA induction. Int J Cancer 19: 317–323

Gahn T, Schildkraut CL (1989) The Epstein-Barr virus origin of plasmid replication, *ori* P, contains both the initiation and termination sites of DNA replication. Cell 58: 527–535

Gerber P (1972) Activation of Epstein-Barr virus by 5-bromodeoxyuridine in "virus-tree" human cells. Proc Natl Acad Sci USA 69: 83–85

Glassy MC, Handley H, Hagiwara H, Royston I (1983) UC-729-6, a human lymphoblastoid B-cell line useful for generating antibody-secreting human–human hybridomas. Proc Natl Acad Sci USA 80: 6327–6331
Gluzman Y (1981) SV40-transformed simian cells support the replication of early SV40 mutants. Cell 23: 175–182
Gluzman Y (ed) (1982) Eurkaryotic viral vectors. Cold Spring Harbor Laboratories, New York
Gluzman Y, Hughes SH (eds) (1988) Eukaryotic viral vectors. Cold Spring Harbor Laboratories, New York
Grogan E, Jensen H, Countryman J, Heston L, Gradoville L, Miller G (1987) Transfection of a rearranged viral DNA fragment, WZhet, stably converts latent Epstein-Barr viral infection to productive infection in lymphoid cells. Proc Natl Acad Sci USA 84: 1332–1336
Gussander E, Adams A (1984) Electron microscopic evidence for replication of circular Epstein-Barr virus genomes in latently infected Raji cells. J Virol 52: 549–556
Haase SB, Heinzel SS, Krysan PJ, Calos MP (1989) Improved EBV shuttle vectors. Mutat Res 220: 125–132
Hambor JE, Hauer CA, Shu H-K, Groger RK, Kaplan DR, Tykocinski ML (1988) Use of an Epstein-Barr virus episomal replicon for anti-sense RNA-mediated gene inhibition in a human cytotoxic T-cell clone. Proc Natl Acad Sci USA 85: 4010–4014
Hammarskjold M-L, Wang S-C, Klein G (1986) High-level expression of the Epstein-Barr virus EBNA1 protein in CV1 cells and human lymphoid cells using a SV40 late replacement vector. Gene 43: 41–50
Hammerschmidt W, Sugden B (1988) Identification and characterization of *ori* lyt, a lytic origin of DNA replication of Epstein-Barr virus. Cell 55: 427–433
Hampar B, Derge JG, Martos LM, Walker JL (1972) Synthesis of Epstein-Barr virus after activation of the viral genome in a "virus-negative" human lymphoblastoid cell (Raji) made resistant to 5-bromodeoxyuridine. Proc Natal Acad Sci USA 69: 78–82
Hampar B, Tanaka A, Nonoyama M, Derge J (1974) Replication of the resident repressed Epstein-Barr virus genome during the early S phase (S-1 period) of nonproducer Raji cells. Proc Natl Acad Sci USA 71: 631–633
Hanahan D (1983) Studies on transformation of *Escherichia coli* with plasmids. J Mol Biol 166: 557–580
Heinzel SS, Krysan PJ, Calos MP, DuBridge RB (1988) Use of simian virus 40 replication to amplify Epstein-Barr virus shuttle vectors in human cells. J Virol 62: 3738–3746
Heller R, Song K, Villaret D, Margolskee RF, Dunne J, Hayakawa H, Ringold GM (1990) Amplified expression of tumor necrosis factor receptor in cells transfected with Epstein-Barr virus shuttle vector cDNA libraries. J Biol Chem 264: 5708–5717
Henle W, Henle G (1972) The relation of the Epstein-Barr virus to Burkitt's lymphoma. In: Biggs RM, de-The G, Payne LN (eds) Oncogenesis and herpesviruses. IARC, Lyon
Henle W, Henle G, Zajac BA, Pearson G, Waubke R, Scriba M (1970) Differential reactivity of human serums with early antigens induced by Epstein-Barr virus, Science 169: 188–190
Henle W, Diehl V, Kohn G, ZurHausen H, Henle G (1987) Herpes-type virus and chromosome marker in normal leukocytes after growth with irradiated Burkitt cells. Science 157: 1064–1065
Hennessy K, Kieff E (1983) One of two Epstein-Barr virus nuclear antigens contains a glycine-alanine copolymer domain. Proc Natl Acad Sci USA 80: 5665–5669
Hennessy K, Kieff E (1985) A second nuclear protein is encoded by Epstein-Barr virus in latent infection. Science 277: 1238–1240
Hennessy K, Heller M, van Santen V, Kieff E (1983) Simple repeat array in Epstein-Barr virus DNA encodes part of the Epstein-Barr nuclear antigen. Science 220: 1396–1398
Hennessy K, Fennewald S, Kieff E (1985) A third viral nuclear protein in lymphoblasts immortalized by Epstein-Barr virus. Proc Natl Acad Sci USA 82: 5944–5948
Hennessy K, Wang F, Bushman EW, Kieff E (1986) Definitive identification of a member of the Epstein-Barr virus nuclear protein 3 family. Proc Natl Acad Sci USA 83: 5693–5697
Hieter P, Mann C, Snyder M, Davis RW (1985) Mitotic stability of yeast chromosomes: a colony color assay that measures nondisjunction and chromosome loss. Cell 40: 381–392
Hirt A (1967) Selective extraction of polyoma DNA from infected mouse cell cultures. J Mol Biol 26: 365–369
Hsiung N, Fitts R, Wilson S, Milne A, Hamer D (1984) Efficient production of hepatitis B surface antigen using a bovine papilloma virus-metallothionein vectors. J Mol Appl Genet 2: 497–506
Hudson GS, Bankier AT, Satchwell SC, Barrell BG (1985) The short unique region of the B95-8 Epstein-Barr virus genome. Virology 147: 81–98

Jalanko A, Kallio A, Salminen M, Ulmanen I (1989) Efficient synthesis of influenza virus hemagglutinin in mammalian cells with an extrachromosomal Epstein-Barr virus vector. Gene 78: 287–296
Jenson HB, Farrell PJ, Miller G (987) Sequences of the Epstein-Barr virus (EBV) large internal repeat form the center of a 16-kilobase-pair palindrome of EBV (P3HR-1) heterogeneous DNA. J Virol 61: 1495–1506
Kadesch T, Berg, P (1986) Effects of the position of the simian virus 40 enhancer on expression of multiple transcription units in a single plasmid. Mol Cell Biol 6: 2593–2601
Kallin B, Dillner J, Ernberg I, Ehlin-Hendriksson B, Rosen A, Henle W, Henle G, Klein G (1986) Four virally determined nuclear antigens are expressed in Epstein-Barr virus-transformed cells. Proc Natl Acad Sci USA 83: 1499–1503
Kavathas P, Sukhatme V, Herzenberg LA, Parnes JR (1984) Isolation of the gene encoding the human T-lymphocyte differentiation antigen Leu-2 (T8) by gene transfer and cDNA subtraction. Proc Natl Acad Sci USA 81: 7688–7692
Kintner CR, Sugden B (1979) The structure of the termini of the DNA of Epstein-Barr virus. Cell 17: 661–671
Kioussis D, Wilson F, Daniels C, Leveton C, Taverne J, Playfiar JHL (1987) Expression and rescuing of a cloned human tumor necrosis factor gene using an EBV-based shuttle cosmid vector. EMBO J 6: 355–361
Klein G (1979) The relationship of the virus to nasopharyngeal carcinomas. In: Epstein MA, Achong BG (eds) The Epstein-Barr virus. Springer, Berlin Heidelberg New York, pp 339–350
Krysan P, Haase S, Calos P (1989) Isolation of human sequences that replicate autonomously in human cells. Mol Cell Biol 9: 1026–1033
Laux G, Freese UK, Fischer R, Polack A, Kofler E, Bornkamm GW (1988) TPA-inducible Epstein-Barr virus genes in Raji cells and their regulation. Virology 162: 503–507
Lebkowski JS, Clancy S, Calos MP (1985) Simian virus 40 replication in adenovirus-transformed human cells antagonizes gene expression. Nature 317: 169–171
Lehn H, Sauer G (1984) The physical state of hybrid genones containing simian virus 40 and bovine papilloma virus DNA sequences in transformed clonal cell lines. J Gen Virol 65: 1413–1418
Leyvraz S, Henle W, Chahinian AP, Perlmann C, Klein G, Gordon RE, Rosenblum M, Holland JF (1985) Association of Epstein-Barr virus with thymic carcinoma. N Engl J Med 312: 1296–1299
Lindahl T, Adams A, Bjursell G, Bornkamm GW, Kaschka-Dierich B, Jehn V (1976) Covalently closed circular duplex DNA of Epstein-Barr virus in a human lymphoid cell line. J Mol Biol 102: 511–530
Littman DR, Thomas Y, Maddon PJ, Chess L, Axel R (1985) The isolation and sequence of the gene encoding T8: a molecule defining functional classes of T lymphocytes. Cell 40: 237–246
Lombardi L, Newcomb EW, Della-Favera R (1987) Pathogenesis of Burkitt lymphoma: expression of an activated c-myc oncogene causes the tumorigenic conversion of EBV-infected human B-lymphocytes. Cell 49: 161–170
Luka J, Kallin B, Klein G (1979) Induction of the Epstein Barr virus (EBV) cycle in latently infected cells by n-butyrate. Virology 94: 228–231
Lupton S, Levine AJ (1985) Mapping genetic elements of Epstein-Barr virus that facilitate extrachromosomal persistence of Epstein-Barr virus-derived plasmids in human cell. Mol Cell Biol 5: 2533–2542
Lutfalla G, Armbruster L, Dequin S, Bertolotti R (1989) Construction of an EBNA-producing line of well differentiated human hepatoma cells and of appropriate Epstein-Barr virus based shuttle vectors. Gene 76: 27–39
Margolskee RF, Nathans D (1984) Simian virus 40 mutant T antigens with relaxed specificity for the nucleotide sequence at the viral DNA origin of replication. J Virol 49: 386–393
Margolskee RF, Kavathas P, Berg P (1988) Epstein-Barr virus shuttle vector for stable episomal replication of cDNA expression libraries in human cells. Mol Cell Biol 8: 2837–2847
Mecsas J, Sugden B (1987) Replication of plasmids derived from bovine papilloma virus type 1 and Epstein-Barr virus in cells in culture. Annu Rev Cell Biol 3: 87–108
Miller G (1985) Epstein-Barr virus. In: Fields BN (ed) Virology. Raven, New York, pp 563–589
Mitrani-Rosenbaum S, Maroteaux L, Mory Y, Revel M, Howley P (1983) Inducible expression of the human interferon β_1, gene linked to a bovine papilloma virus DNA vector and maintained extrachromosomally in mouse cells. Mol Cell Biol 3: 233–240
Nonoyama M, Pagano JS (1972) Separation of Epstein-Barr virus DNA from large chromosomal DNA in non-virus producing cells. Nature New Biol 238: 169–171

Oguro MO, Shimizu N, Ono Y, Takada K (1987) Both the rightward and the leftward open reading frames within the *Bam*HI M DNA fragment of Epstein-Barr virus act as *trans*-activators of gene expression. J Virol 61: 3310–3313
Okayama H, Berg P (1982) High-efficiency cloning of full-length cDNA. Mol Cell Biol 2: 161–170
Okayama H, Berg P (1983) A cDNA cloning vector that permits expression of cDNA inserts in mammalian cell. Mol Cell Biol 3: 280–289
Okayama H, Berg P (1985) Bacteriophage lambda vector for transducing cDNA clone library into mammalian cell. Mol Cell Biol 5: 1136–1142
Pattengale PK, Smith RW, Gerber P (1973) Selective transformation of B lymphocytes by EB virus. Lancet ii: 93–94
Pintel D, Merchlinsky MJ, Ward DC (1984) Expression of minute virus of mice structural proteins in murine cell lines transformed by bovine papillomavirus-minute virus of mice plasmid chimera. J Virol 52: 320–327
Pope JH, Horne MK, Scott W (1968) Transformation of fetal human leukocytes in vitro by filtrates of a human leukaemic cell line containing herpes-like virus. Int J Cancer 3: 857–866
Potter H, Weir L, Leder P (1984) Enchancer-dependent expression of human κ immunoglobulin genes introduced into mouse pre-B lymphocytes by electroporation. Proc Natl Acad Sci USA 81: 7161–7165
Purtillo DT, Bhawan J, DeNicola L, Hutt LM, Szymanski I, Yang JPS, Boto W, Maier R, Thorley-Lawson D (1978) Epstein-Barr virus infections in the X-linked recessive lymphoproliferative syndrome. Lancet i: 798–802
Rawlins DR, Milman G, Hayward SD, Hayward GF (1985) Sequence-specific DNA binding of the Epstein-Barr virus nuclear antigen (EBNA-1) to clustered sites in the plasmid maintenance region. Cell 42: 859–868
Razzaque A, Mizusuawa H, Seidman MM (1983) Rearrangement and mutagenesis of a shuttle vector plasmid after passage in mammalian cells. Proc Natl Acad Sci USA 80: 3010–3014
Reedman BM, Klein G (1973) Cellular localization of an Epstein-Barr virus (EBV) associated complement-fixing antigen in producer and non-producer lymphoblastoid cell lines. Int J Cancer 11: 499–520
Reisman D, Sugden B (1986) *trans* Activation of an Epstein-Barr viral transcriptional enchancer by the Epstein-Barr viral nuclear antigen 1. Mol Cell Biol 6: 3838–3846
Reisman D, Yates J, Sugden B (1985) A putative origin of replication of plasmids derived from Epstein-Barr virus is composed of two *cis*-acting components. Mol Cell Biol 5: 1822–1832
Rooney CM, Rowe DT, Ragot T, Farrell PF (1989) The spliced BZLF1 gene a Epstein-Barr virus (EBV) transactivates an early EBV promoter and induces the virus productive cycle. J Virol 63: 3109–3116
Shimizu N, Yamaki M, Sakuma S, Ono Y, Takada K (1988) Three Epstein-Barr virus (EBV)-determined nuclear antigens induced by the BamHI E region of EBV DNA. Int J Cancer 41: 744–751
Shimizu Y, Koller B, Geraghty D, Orr H, Shaw S, Kavathas P, DeMars R (1986) Transfer of cloned human class I major histocompatibility complex genes into HLA mutant human lymphoblastoid cells. Mol Cell Biol 6: 1074–1087
Shortle D, Nathans D (1978) Local mutagenesis: a method for generating viral mutants with base substitutions in pre-selected regions of the viral genome. Proc Natl Acad Sci USA 75: 2170–2174
Shortle D, Nathan D (1979) Regulatory mutants of simian virus 40: constructed mutants with base substitutions at the origin of DNA replication. J Mol Biol 131: 801–817
Shortle D, Margolskee RF, Nathans D (1979) Mutational analysis of the simian virus 40 replicon: pseudorevertants of mutants with defective replication origin. Proc Natl Acad Sci USA 76: 6128–6131
Skare J, Strominger JL (1980) Cloning and mapping of Bam HI endonuclease fragments of DNA from the transforming B95-8 strain of Epstein-Barr virus. Proc Natl Acad Sci USA 77: 3860–3864
Skare J, Farley J, Strominger JL, Fresen KO, Cho MS, zur Hausen H (1985) Transformation by Epstein-Barr virus requires DNA sequences in the region of *Bam*HI fragements Y and H. J Virol 55: 286–297
Southern PJ, Berg P (1982) Transformation of mammalian cells to antibiotic resistance with a bacterial gene under control of the SV40 early region promoter. J Mol Appl Genet 1: 327–341
Speck SH, Strominger JL (1985) Analysis of the transcript encoding the latent Epstein-Barr virus nuclear antigen I: a potentially polycistronic message generated by long-range splicing of several exons. Proc Natl Acad Sci USA 82: 8305–8309
Speck SH, Strominger JL (1987) Epstein-Barr virus transformation. Prog Nucleic Acid Res Mol Biol 34: 189–207

Strand BC, Schuster TC, Hopkins RF, Neubauer RH, Rabin H (1981) Identification of an Epstein-Barr virus nuclear antigen by fluoroimmunoelectrophoresis and radioimmunoelectrophoresis. J Virol 38: 996–1004

Sugden B (1984) Expression of virus-associated functions in cells transformed in vitro by Epstein-Barr virus: Epstein-Barr virus cell surface antigen and virus-release from transformed cells. In: Purtillo DT (ed) Immune deficiency and cancer, Epstein-Barr virus and lymphoproliferative malignancies. Plenum, New York, p 165

Sugden B, Warren N (1988) Plasmid origin of replication of Epstein-Barr virus, *ori* P, does not limit replication in *cis*. Mol Biol Med 5: 85–94

Sugden B, Warren N (1989) A promoter of Epstein-Barr virus that can function during latent infection can be transactivated by EBNA-1, a viral protein required for viral DNA replication during latent infection. J Virol 63: 2644–2649

Sugden B, Phelps M, Domoradzki J (1979) Epstein-Barr virus DNA is amplified in transformed lymphocytes. J Virol 31: 590–595

Sugden B, Marsh K, Yates J (1985) A vector that replicates as a plasmid and can be efficiently selected in B-lymphoblasts transformed by Epstein-Barr virus. Mol Cell Biol 5: 410–413

Summers WP, Grogan EA, Shedd D, Robert M, Lieu C-R, Miller G (1982) Stable expression in mouse cells of nuclear neoantigen after transfer of a 3.4 megadalton cloned fragment of Epstein-Barr virus DNA. Proc Natl Acad Sci USA 79: 5688–5692

Takada K (1984) Cross-linking of cell surface immunoglobulins induces Epstein-Barr virus in Burkitt lymphoma lines. Int J Cancer 33: 27–32

Takada K, zur Hausen H (1984) Induction of Epstein-Barr virus antigens by tumor promoters for epidermal and nonrepidermal tissues. Int J Cancer 33: 491–496

Takada K, Shimizu N, Ogura M, Oni Y (1986a) Identification of coding regions for various Epstein-Barr virus-specific antigens by gene transfer and serology. J Virol 60: 324–330

Takada K, Shimizu N, Sakuma S, Ono Y (1986b) *trans* Activation of the latent Epstein-Barr virus (EBV) genome after transfection of the EBV DNA fragment. J Virol 57: 1016–102

Tovy MG, Lenoir G, Begon-Lours J (1978) Activation of latent Epstein-Barr virus by antibody to human IgM. Nature 276: 270–272

Tykocinski ML, Shu H-K, Ayers DJ, Walter EI, Getty RR, Groger RK, Hauer CA, Medof ME (1988) Glycolipid reanchoring of T-lymphocyte surface antigen CD8 using the 3′ end sequence of decay-accelerating factor's mRNA. Proc Natl Acad Sci USA 85: 3555–3559

Wang Y, Stratowa C, Schaefer-Ridder M, Doehmer J, Hofschneider PH (1983) Enhanced production of hepatitis B surface antigen in NIH 3T3 mouse fibroblasts by using extrachromosomally replicating bovine papillomavirus vector. Mol Cell Biol 3: 1032–1039

Weaver JF, McCormick F, Manos MM (1988) Production of recombinant human CSF-1 in an inducible mammalian expression system. Biotechnology 6: 287–290

Wysokenski D, Yates J (1989) Multiple EBNA-1 binding sites are required to form an EBNA-1 dependent enhancer and to activate a minimal replicative origin within *ori* P of Epstein-Barr virus. J Virol 63: 2657–2666

Yates JL, Warren N, Reisman D, Sugden B (1984) A *cis*-acting element from the Epstein-Barr viral genome that permits stable replication of recombinant plasmids in latently infected cells. Proc Natl Acad Sci USA 81: 3806–3810

Yates, JL, Warren N, Sugden B (1985) Stable replication of plasmids derived from Epstein-Barr virus in various mammalian cells. Nature 313: 812–815

Young JM, Cheadle C, Foulke JS Jr, Drohan WN, Sarver N (1988) Utilization of an Epstein-Barr virus replicon as a eurkaryotic expression vector. Gene 62: 171–185

Young RA, Davis R (1983) Efficient isolation of genes by using antibody probes. Proc Natl Acad Sci USA 80: 1194–1198

Zinn K, DiMaio D, Maniatis T (1983) Identification of two distinct regulatory regions adjacent to the human β-interferon gene. Cell 34: 865–879

zur Hausen H, O'Neill FJ, Freese UK, Hecker E (1978) Persisting oncogenic herpesvirus induced by the tumor promoter TPA. Nature 272: 373–375

Use of Adeno-Associated Virus as a General Transduction Vector for Mammalian Cells

N. MUZYCZKA

1 The Life Cycle of Adeno-Associated Virus

Adeno-associated virus (AAV) is a human virus that can be propagated either as an integrated provirus or by lytic infection (ATCHISON et al. 1965; HOGGAN et al. 1966, 1972). The ability to form a latent infection appears to be an integral part of the AAV life cycle. Except under special circumstances (YACOBSON et al. 1987; SCHLEHOFER et al. 1986; YALKINOGLU et al. 1988), AAV requires the presence of a helper virus to initiate a productive viral infection (Fig. 1). Members

Department of Microbiology, SUNY Stony Brook Medical School, Stony Brook, NY 11794, USA

Current Topics in Microbiology and Immunology, Vol. 158

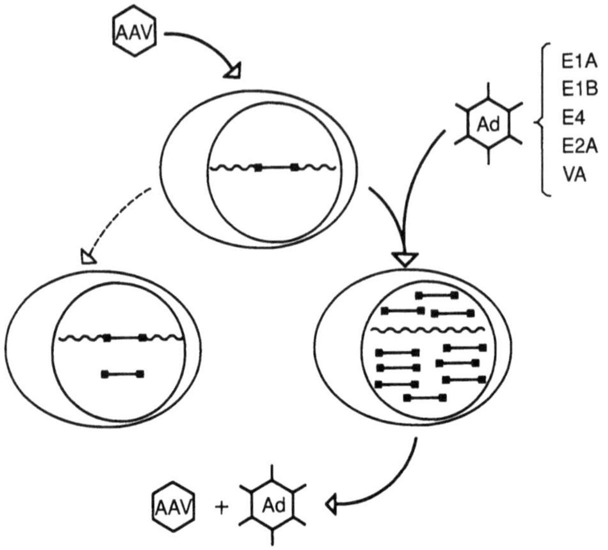

Fig. 1. AAV life cycle. When cells are infected with AAV in the absence of helper virus, the AAV genome integrates into a host chromosome. Superinfection of the proviral line with adenovirus rescues the provirus and induces amplification of the AAV genome. The result is a viral stock containing a mixture of AAV and adenovirus. The adenovirus genes required for helper function are indicated on the *right*. The pathway indicated by a *dotted line* illustrates the fact that under some conditions the AAV genome can excise and replicate in the absence of helper virus. The *wavy line* represents the host chromosome; *solid lines* and *solid squares* represent the AAV genome and the AAV terminal repeats

of either the herpes or adenovirus families can provide the necessary helper functions (ATCHISON et al. 1965; MELNICK et al. 1965; HOGGAN et al. 1966; BULLER et al. 1981; MCPHERSON et al. 1985) and vaccinia virus can provide at least partial helper function (SCHLEHOFER et al. 1986). In the absence of a helper virus AAV produces no progeny virus, but instead integrates into a host chromosome (HOGGAN et al. 1972; BERNS et al. 1975; HANDA et al. 1977; CHEUNG et al. 1980). With rare exceptions, AAV proviruses appear to be stable. However, if a cell line that is carrying an AAV provirus (Fig. 1) is subsequently superinfected with a helper virus, the AAV genome is excised and proceeds through a normal productive infection (HOGGAN et al. 1972; CHEUNG et al. 1980). This ability to establish a latent infection which can later be rescued appears to be a mechanism for ensuring the survival of AAV in the absence of a helper virus. The unusual life cycle of AAV has led a number of parvovirus laboratories to explore the possibility of using AAV as a general mammalian transduction vector. This review will focus on the progress that has been made in using AAV as a vector and, also, point out the problems that still remain. For more extensive information on AAV biology, the reader should consider several recent reviews (BERNS and BOHENZKY 1987; BERNS 1990; PATTISON 1990).

2 Genetics of AAV

Because AAV relies on a second virus for productive growth and does not itself produce virus plaques, it has been virtually impossible to do a conventional genetic analysis of AAV. This problem was essentially solved when SAMULSKI et al. (1982) discovered that prokaryotic clones of wild type AAV are infectious. If form I AAV plasmid DNA is transfected into human cells in the presence of adenovirus, the AAV sequences are rescued from the plasmid and proceed through a normal productive infection (SAMULSKI et al. 1982). Aside from providing a model for studying the phenomenon of rescue, the isolation of infectious AAV plasmids has provided an easy method for constructing mutants within the AAV genome as well as the AAV-based vectors described below. In addition to the original infectious clone (pSM620) isolated by SAMULSKI et al. (1982), two other wild type AAV clones have been isolated, pAV1 (LAUGHLIN et al. 1983) and *sub*201 (SAMULSKI et al. 1987). All three clones are infectious but *sub*201, which is missing the first 13 base pairs (bp) from both ends of the AAV genome, has a somewhat lower infectivity because it is apparently partially defective for rescue.

2.1 Capsid Structure

AAV is a member of the parvovirus family (SIEGEL et al. 1985) which has been divided into the autonomous and dependovirus (or adeno-associated virus) groups. In contrast to AAV, the autonomous parvoviruses do not require helper virus coinfection for productive growth. Five serotypes of AAV have been identified (1–5) but the most extensively characterized is AAV-2. AAV-2 virions contain a single-stranded linear DNA molecule which is approximately 4.7 kilobases (kb) long (SRIVASTAVA et al. 1983). Both strands of AAV DNA are packaged with equal efficiency (ROSE et al. 1969; MAYOR et al. 1969; BERNS and ROSE 1970; BERNS and ADLER 1972) and both are equally infectious (SAMULSKI et al. 1987). The nonenveloped virion particle is icosohedral in shape and one of the smallest that has been described, about 20–24 nm in diameter (HOGGAN 1970; MELNICK et al. 1965; TSAO et al. 1991) with a density of 1.41 g/cm^3 (DE LA MAZA and CARTER 1980a, b). The relatively high density of AAV particles allows them to be easily separated by CsCl density centrifugation from adenovirus helper particles whose density is approximately 1.35 g/cm^3 (DE LA MAZA and CARTER 1980a). AAV particles are resistant to heat (56° C for 1 h) and low pH and moderately resistant to detergents and some proteases (BACHMANN et al. 1979). The virion is composed of three proteins, VP1, VP2, and VP3 (ROSE et al. 1971; JOHNSON et al. 1971, 1975, 1977), which are approximately 87, 73, and 61 kDa respectively. VP3 represents 90% of the total virion protein; VP2 and VP1 account for approximately 5% each. All three of the capsid proteins are N-acetylated (BECERRA et al. 1985). As yet, little is known about the organization of these proteins in the virion. However, TSAO et al. (1991) have recently reported information about the crystalline structure of canine parvovirus and, presumably, the structure of AAV virus particles is similar.

2.2 Genome Structure

The complete nucleotide sequence of AAV-2 has been reported by SRIVASTAVA et al. (1983). Two corrections to the original AAV-2 sequence (SRIVASTAVA et al. 1983; Genbank locus AA2CG) have been made: (1) the sequence GGGCCC is present between nucleotides 3759 and 3760 of the original sequence (R. WRIGHT and N. MUZYCZKA, unpublished), and (2) a T was incorrectly inserted in the original sequence at nucleotide 2429 (R. LOFTS and R.A. MCPHERSON, cited in TREMPE and CARTER, 1988b and BECERRA et al. 1988). The complete sequence contains 4680 bases.

The AAV genome contains a terminal inverted repeat (KOCZOT et al. 1973; GERRY et al. 1973; BERNS and KELLY, 1974; LUSBY et al. 1980) which is 145 bp long (Fig. 2). The first 125 bases of the inverted repeat can form a T shaped hairpin structure (LUSBY et al. 1980) which is composed of two small internal palindromes (Figs. 2B and C) flanked by a larger palindrome (A). The terminal sequence can exist in one of two configurations: flip (in which the B palindrome is closer to the 3′ end) or flop (in which the C palindrome is closer to the 3′ end). Both ends of AAV in any particular molecule have an equal probability of being in either the flip or flop configuration (LUSBY et al. 1981), which suggests that inversion of the terminal sequence at one end is independent of events at the other end. In the single-stranded form, the terminal palindrome readily adopts the T-shaped hairpin configuration shown in Fig. 2 as well as several other related configurations (LUSBY et al. 1980).

The internal nonrepeated portion of the genome is divided in half by two large open reading frames (Fig. 3) with maximum coding capacities of 71 kDa (left) and 78 kDa (right). Three promoters have been found and they have been

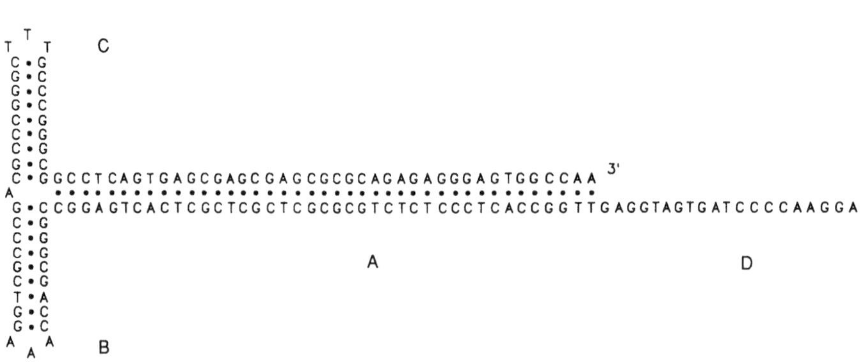

Fig. 2. The AAV inverted terminal repeat. See text of details

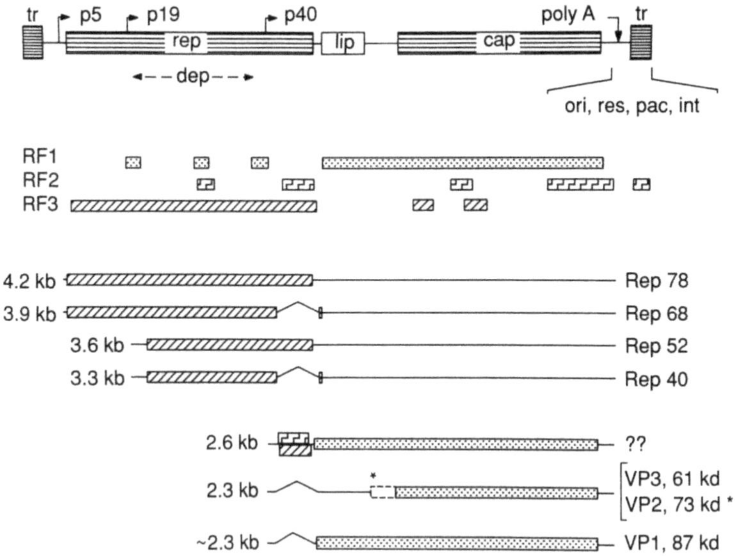

Fig. 3. AAV genome. *Top*, genetic map of AAV showing the regions defined by mutations: rep, lip, cap, and the terminal repeats (*tr*). *ori*, *res*, *pac*, and *int* indicate a segment of the genome that is sufficient for replication, rescue, packaging, and integration of AAV DNA. Also shown is the polyadenylation site (*bent downward arrow*) and the three AAV promoters (*bent arrows*). *Below*, the major open reading frames in each frame and the known AAV mRNAs. The size of each mRNA is shown on the *left*; the proteins synthesized from each mRNA are shown on the *right*; the reading frames used in each mRNA are also shown. The *double question mark* indicates that no known function exists for this message. The *asterisk* indicates that VP2 is synthesized from an extended VP3 open reading frame by using an ACG start codon. See text for more details

named according to their approximate map positions: p5, p19, and p40 (LUSBY and BERNS 1982; LAUGHLIN et al. 1979; GREEN and ROEDER 1980ab; GREEN et al. 1980). Messenger RNA's initiated at these three promoters all terminate at the same polyadenylation site at map position 95 (SRIVASTAVA et al. 1983; LAUGHLIN et al. 1979; GREEN and ROEDER 1980a). In addition, the AAV transcripts share a common intron (nucleotides 2107–2227) and in each case both the spliced and unspliced mRNA species are detected in infected cells (LAUGHLIN et al. 1979; GREEN and ROEDER 1980ab). Finally, at least one alternately spliced p40 mRNA has recently been identified (TREMPE and CARTER 1988b; CASSINOTTI et al. 1989). The new message uses the same donor splice site as the major intron (nucleotide 2107) but a different acceptor site (nucleotide 2201). The resulting message is only slightly larger than the previously identified 2.3 kb mRNA and it allows for the expression of the entire right-hand open reading frame (Fig. 3).

2.3 Rep Mutants

Mutations in the left-hand open reading frame were originally characterized (HERMONAT et al. 1984; TRATSCHIN et al. 1984a) as defective for viral DNA replication (rep^-). They were subsequently found to be pleiotropic mutations that are also defective for the transactivation of AAV promoters (LABOW et al. 1986; TRATSCHIN et al. 1986), for the repression of heterologous gene expression (LABOW et al. 1987; TREMPE and CARTER 1988a; ANTONI et al. 1991), and under some conditons for the repression of AAV-coded gene expression (TRATSCHIN et al. 1986; BEATON et al. 1989; CHEJANOVSKY and CARTER 1990). The Rep protein may also be responsible for the anti-oncogenic effect of AAV infections. Infection with AAV reduces the frequency of in vitro cell transformation and tumor formation induced by adenovirus and herpes virus (OSTROVE et al. 1981; KIRSCHTEIN et al. 1968; MAYOR et al. 1973, 1976; CUKOR et al. 1975; DE LA MAZA and CARTER 1981; CASTO and GOODHEART 1972) and recently the Rep protein has been shown to be necessary for the suppression of transformation by bovine papilloma virus, SV40 DNA, the human, bladder cancer derived *ras* gene, and a combination of the activated human *ras* gene and the adenovirus E1A gene (HERMONAT, 1989, 1991; KHLEIF et al. 1991).

A family of at least four viral proteins are synthesized from the *rep* region (Fig. 2) and these have been named according to their apparent molecular weight (MENDELSON et al. 1986; TREMPE et al. 1987; SRIVASTAVA et al. 1983). The two mRNAs initiated from the p5 promoter are used for the synthesis of Rep 78 and Rep 68. Rep 52 and 40 are translated from the p19 messages and their amino acid sequences are identical to the C-terminal sequences of Rep 78 and 68, respectively. In addition two smaller Rep proteins have been identified in AAV infected cells (TREMPE et al. 1987) and these are believed to be the result of translation initiation at internal ATG codons in the p19 mRNA. Three of the Rep proteins (Rep 68, 78, and 40) have been found to consist of two species of slightly different molecular weight in the monomer form (IM and MUZYCZKA 1990; REDEMAN et al. 1989). The reason for the heterogeneity of these three proteins is not known.

Mutations which map exclusively within the Rep 78 and 68 coding sequence display the full *rep* phenotype (HERMONAT et al. 1984; LABOW et al. 1986; TRATSCHIN et al. 1986; LABOW et al. 1987). As yet, it is not certain what role is played by the two p19 Rep proteins during a productive viral infection. Recent genetic and biochemical evidence suggests that they are not essential for duplex viral DNA replication (CHEJANOVSKY and CARTER 1989b; SNYDER et al. 1990b) but may contribute to the ability of AAV to repress expression from heterologous promoters (LABOW et al. 1987; MENDELSON et al. 1988b) and may be necessary for generating single-stranded progeny during AAV DNA replication (CHEJANOVSKY and CARTER, 1989b). It also is not clear why both spliced and unspliced versions of the Rep proteins are produced. Mutants within the intron region have quite variable phenotypes. One frameshift insertion mutant (*ins*42) reduced DNA replication 100 fold and trans-activation of the p40 promoter 5 fold (HERMONAT et al. 1984; LABOW et al. 1987). A small deletion mutant (p*dl*204) also reduced DNA replication (SAMULSKI et al. 1987), but larger deletions (e.g., pAV*dl*HiX2) that included the intron region were viable for DNA replication (TRATSCHIN et al. 1984a). Finally, one insertion mutation just upstream of the splice donor site was viable for both DNA replication and *trans*-activation (TRATSCHIN et al. 1984b; TREMPE and CARTER 1988a) when the insert was in one orientation (pTS1) but not in the opposite orientation (pAVHiTac). Interpretation of these results is further complicated by the fact that there are two small open reading frames entirely within the intron which could be translated from the 2.6 kb unspliced p40 messenger RNA (Fig. 2) to produce proteins about 10 kDa in size. Although one or both of these proteins has been detected by in vitro translation assays, their significance is unknown (BECERRA et al. 1988).

One class of mutants in the *rep* region has a *cis* effect on the accumulation of Rep mRNAs. Some of the deletions which map between mu 10 and 37 accumulate the p5 and p19 transcripts to abnormally high levels (LABOW et al. 1986; TREMPE and CARTER 1988a). This suggests that there is an attenuator sequence within this region (Fig. 2, *dep*) that modulates the level of Rep mRNAs, but the exact location of this sequence and its mechanism are unknown.

2.4 Capsid Mutants

As in the case of the Rep proteins, the right-hand open reading frame of AAV codes for a family of proteins (Fig. 3). The amino acid sequence of the major capsid protein VP3 is contained within the two larger and less abundant capsid proteins VP1 and VP2 (JOHNSON et al. 1977; JANIK et al. 1984; MCPHERSON and ROSE 1983). VP3 is synthesized from the 2.3-kb mRNA (Fig. 3) by using an ATG at nucleotide 2809 (BECERRA et al. 1985, 1988). Mutations within the VP3 coding region make no capsid proteins (cap^-) and characteristically accumulate little if any single-stranded AAV progeny DNA (HERMONAT et al. 1984; TRATSCHIN et al. 1984a). It is known that AAV synthesizes empty procapsids prior to packaging (MYERS and CARTER 1980). Therefore, the most likely explanation for the *cap* phenotype is that, in the absence of procapsids, newly synthesized AAV DNA is rapidly cycled into the replicating pool of double-stranded DNA or is degraded (HERMONAT et al. 1984; TRATSCHIN et al. 1984a).

The second largest AAV capsid protein, VP2, is made from the same 2.3 kb mRNA as VP3 (BECERRA et al. 1985, 1988). This is believed to occur by initiation of translation at a novel eukaryotic start codon which is 65 codons upstream of the VP3 start site. Identification of this codon (ACG) rests primarily on sequencing of VP2 peptide fragments (BECERRA et al. 1985) and, as yet, no mutations specific for VP2 have been isolated.

Mutations within the N-terminal region of the right hand orf (mu 47–55) have a distinct phenotype which has been called *lip* (HERMONAT et al. 1984) or *inf* (TRATSCHIN et al. 1984a). The biochemical defect of *lip/inf* mutants is an altered or missing VP1 protein (JANIK et al. 1984). VP1 is synthesized from the alternately spliced p40 mRNA mentioned earlier (Fig. 3) which allows for the synthesis of VP1 from the start codon at nucleotide 2202 (TREMPE and CARTER 1988b; BECERRA et al. 1988; CASSINOTTI et al. 1989). *lip/inf* mutations produce low yields of infectious particles but are viable for the synthesis of both double-stranded replicating form DNA and single-stranded progeny DNA. This implies that *lip/inf* mutants are making a procapsid and packaging single-stranded AAV DNA but are defective at some later stage of capsid processing. It is not known whether *lip/inf* mutants make particles which are non-infectious, particles which are unstable, or, simply, reduced numbers of particles. MYERS and CARTER (1980) have shown that AAV particle synthesis goes through an intermediate capsid form that contains a full complement of AAV DNA. It has been suggested that the *lip/inf* defect may represent a failure to process this intermediate capsid to the mature AAV viral particle (HERMONAT et al. 1984; TRATSCHIN et al. 1984a) but, as yet, there is no direct evidence for this possibility.

It is not known whether only the p40 initiated messages are capable of alternate splicing or whether all AAV mRNAs occur in two spliced forms. Finally, although the two alternately spliced p40 mRNAs account for the synthesis of all three capsid proteins, it should be mentioned that MCPHERSON and ROSE (1983) have reported that multiple forms of both VP1 and VP3 (but not VP2) are synthesized in vivo and during in vitro translation of AAV mRNA. The latter experiment suggests that the variant proteins may not be due to post-translational modifications. The mechanism that is used to generate the heterogeneity is unknown.

2.5 Multiple Functions of the Terminal Repeats

The terminal repeats are required in *cis* for AAV DNA replication and for rescue or excision from prokaryotic plasmids (SAMULSKI et al. 1983, 1987; SENEPATHY et al. 1984; GOTTLIEB and MUZYCZKA 1988). In addition, the terminal repeats appear to be the minimum sequences required for AAV proviral integration and for packaging of AAV DNA into virions (MCLAUGHLIN et al. 1988; SAMULSKI et al. 1989). In the case of DNA replication it is clear that most of the terminal 125 base palindrome is required for DNA replication and terminal resolution (BOHENZKY et al. 1988; LEFEBVRE et al. 1984; IM and MUZYCZKA 1989; ASHKTORAB and SRIVASTAVA 1989). However, as yet there is no clear indication whether any of the other functions of the terminal repeats (packaging, integration, or rescue) require all or only some of the terminal sequence.

2.6 DNA Replication

The two major replicating forms of AAV DNA are a linear duplex molecule with open ends and a linear duplex in which one of the ends is covalently joined (Fig. 4). This observation led STRAUS et al. (1976a) to suggest that AAV DNA replicates by a mechanism that had been suggested by CAVALIER-SMITH (1974) to solve the problem of how chromosomes replicate their ends. The model was further refined by BERNS and HAUSWIRTH (1979) when the terminal sequence of AAV was determined (LUSBY et al. 1980). It is believed that the terminal

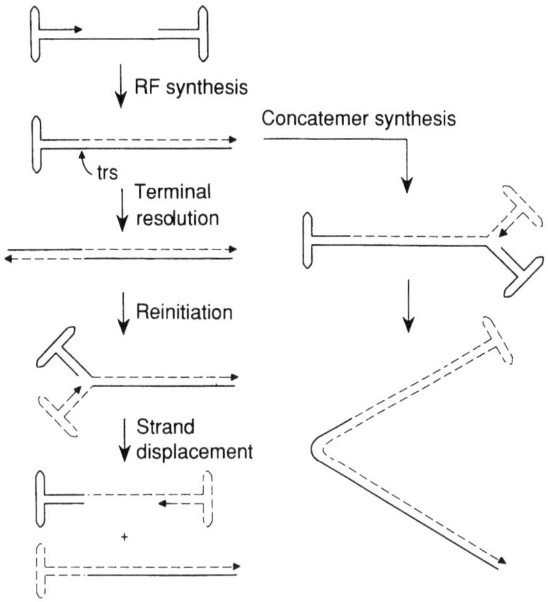

Fig. 4. AAV DNA replication. A simplified diagram of the intermediates formed during AAV DNA replication starting with a single-stranded viral DNA molecule. *Arrows* indicate the 3′ OH ends; *solid lines* represent the parental DNA molecule; *dotted lines* represent new DNA synthesis; *trs* is the terminal resolution site. See text for details

palindrome of AAV acts as a primer for the synthesis of a duplex linear molecule in which one of the ends is covalently joined (Fig. 4). The covalently joined end is then resolved to an open duplex end by a two-step process. First, a nick is made on the parental strand at a terminal resolution site (*trs*) opposite the original 3′ OH primer position. This creates a new 3′ OH primer which then allows the end to be repaired. Following this, the terminal repeat is denatured so that the palindromic sequences can again adopt the hairpin configuration. Finally, the hairpin primer is used for displacement synthesis to generate a new replicative form (RF) molecule and a single-stranded progeny molecule which is packaged.

All of the currently available evidence supports the model. First, as already mentioned, duplex RF molecules in which one end is covalently joined are found in AAV-infected cells (Straus et al. 1976a). Also found are dimer concatemers of the AAV genome. These are presumably the result of a second round of replication on monomer RF molecules which have not been resolved (see Fig. 4). As expected, the dimers contain two copies of the full length AAV genome in an inverted orientation (head to head or tail to tail) (Straus et al. 1976a; Hauswirth and Berns 1979). In addition, spontaneous deletion mutants which are missing internal regions of AAV DNA comprise a major type of AAV defective interfering (DI) genome. These are packaged into virions and are capable of DNA replication. Many of these DI molecules consist of genomes that rapidly reanneal after extraction from virus to form less than unit length linear molecules that are covalently joined at one end (de la Maza and Carter 1980a; Hauswirth and Berns 1979). Second, in vivo pulse chase experiments demonstrate that AAV DNA synthesis is initiated and terminated near the ends of the molecule on each strand (HAUSWIRTH and BERNS, 1977). Third, genetic analysis of deletions in the terminal repeats indicates that they behave like *cis* defective *ori* mutants (SAMULSKI et al. 1983; SENEPATHY et al. 1984) and that both ends of the genome are required. Deletions in one of the terminal repeats are viable but only because they rapidly undergo gene conversion back to the wild type sequence. Mutants with deletions in both termini are completely defective for DNA replication (SAMULSKI et al. 1983; SENEPATHY et al. 1984). The model also predicts that the terminal palindrome should be inverted during resolution and this explains the two orientations of the terminal sequence (flip and flop) that have been found in viral DNA (LUSBY et al. 1980). As expected, the configuration at either end is independent of the other (LUSBY et al. 1981) and recombinant clones that contain the viral terminal repeats in only one orientation generate viral DNA with terminal sequences in both orientations (SAMULSKI et al. 1982). Finally, SNYDER et al. (1990a) have demonstrated that the covalently joined ends of a linear AAV DNA molecule can be resolved correctly in vitro as predicted by the model. The resolution step can be initiated by either Rep 68 (SNYDER et al. 1990b; IM and MUZYCZKA 1990) or Rep 78 (Im and Muzyczka, in preparation). Rep 68, which has been purified to apparent homogeneity, was found to have several biochemical activities that are related to its role in DNA replication. These include the ability to bind to the terminal hairpin DNA (IM and MUZYCZKA 1989; ASHKTORAB and SRIVASTAVA 1989), the ability to cut the appropriate strand of the AAV hairpin at the *trs* site, the ability to covalently attach to the 5′ end of the nick during terminal resolution, and an associated DNA helicase activity (IM and MUZYCZKA 1990). As yet, it is not known whether any of these Rep activities have a role in the transactivation or repressor functions of the *rep* gene.

3 Control of AAV Gene Expression

The control of AAV gene expression is complex; both helper virus and AAV genes are involved. Typically, tissue culture cells that are infected with AAV alone accumulate barely detectable amounts of AAV mRNA or protein and no AAV DNA replication can be detected (ROSE and KOCZOT 1972; LAUGHLIN et al. 1982; REDEMANN et al. 1989; ANTONI et al. 1991). Coinfection with adenovirus initiates a cascade of regulatory events which ultimately result in a high titer of AAV virus and drastically reduced levels of adenovirus production. The initial step in the cascade is the synthesis of the adenovirus E1A gene product which induces transcription from the AAV p5 and p19 promoters and leads to the synthesis of a small amount of the Rep proteins (CHANG et al. 1989; TRATSCHIN et al. 1984b; FLINT and SHENK 1989). One or more of the p5 Rep proteins then induce the synthesis of mRNA from all three AAV promoters in a coordinate fashion to an even higher level (LABOW et al. 1986; TRATSCHIN et al. 1986). Taken together there is an induction of more than 50-fold in the steady state level of

AAV mRNA following adenovirus infection and Rep protein synthesis (LAUGHLIN et al. 1982; LABOW et al. 1986; REDEMANN et al. 1989). In contrast, when adenovirus is not present, the infecting AAV genome is either lost or integrates into host chromosomes, resulting in a relatively low level of expression from all three promoters (LABOW et al. 1986; MENDELSON et al. 1988a; REDEMANN et al. 1989; LAUGHLIN et al. 1982; ANTONI et al. 1991; HOGGAN et al. 1972; ROSE and KOCZOT 1972). Therefore, the E1A–Rep cascade appears to be a decisive transcriptional switch which is used to decide whether AAV will integrate or undergo productive growth.

Relatively little work has been done so far in identifying specific sequence elements that control the three AAV promoters and respond to the Rep protein. AAV mutants in which the terminal repeats have been removed exhibit normal control of AAV gene expression, suggesting that the terminal repeats are not essential for AAV transcription (LABOW et al. 1986; BEATON et al. 1989; TRATSCHIN et al. 1986). However, MCCARTY et al. (1991) have suggested that there may be two redundant mechanisms for Rep induction. They found that in the absence of the terminal repeats, induction of the p40 promoter required the presence of sequences upstream of all three AAV promoters. In contrast, when the terminal repeat was present, Rep could induce the p40 promoter regardless of whether the p5 and p19 promoters were present. They concluded that each AAV promoter contained Rep-responsive sequences. In addition, the terminal repeat was a redundant sequence that could mediate Rep induction. BEATON et al. (1989) have also suggested that the terminal repeat has a role in transcription. They reported that in the presence of the *rep* gene and adenovirus, the terminal sequence could enhance transcription from a nearby SV40 promoter.

If the adenovirus E1A and the AAV *rep* gene products were all that were needed for AAV gene expression, then one would expect adenovirus-transformed human cells which constitutively express the E1A gene products (such as the 293 cell line) to be permissive for AAV virus production. In fact, the adenovirus-transformed 293 cell line is only semipermissive. Infection of 293 cells with AAV produces barely detectable levels of AAV virus (McLaughlin and Muzyczka, in preparation). The semipermissiveness of 293 cells is apparently due to the fact that several other adenovirus gene products are also required for efficient AAV gene expression, either by promoting AAV mRNA accumulation (E1B and E4 regions) (SAMULSKI and SHENK 1988; LAUGHLIN et al. 1982; CARTER et al. 1983) or by promoting AAV mRNA splicing and translation (E2A and VA regions) (WEST et al. 1987; MCPHERSON et al. 1982; MYERS et al. 1980; JANIK et al. 1981; JAY et al. 1981; RICHARDSON and WESTPHAL 1981, 1984; HUANG and HEARING 1989; STRAUS et al. 1976b).

3.1 Repression of Gene Expression by the Rep Protein

In the absence of adenovirus, the Rep protein appears to be capable of repressing both AAV and heterologous gene expression. This was first reported in the case of a *cat* gene that had been put under the control of the AAV p40 promoter (TRATSCHIN et al. 1986; TREMPE and CARTER 1988a; WEST et al. 1987). When transient expression of the *cat* gene was measured in 293 cells in presence and absence of *rep*, it was found that the *rep* gene reduced the amount of chloramphenicol acetyltransferase synthesized even though it increased the level of p40-*cat* mRNA. This led to the suggestion that *rep* repression occurred primarily at the translational level of the p40 mRNA (TREMPE and CARTER 1988a). More recent experiments have suggested that Rep also can repress *cat* gene expression when the *cat* gene is under the control of the AAV p5 and p19 promoters and that adenovirus infection can overcome the repression (BEATON et al. 1989). In this case it was not clear whether repression occurred at the level of transcription or translation. Further support for the possibility that the Rep protein autoregulates its own synthesis comes from the phenotype of a *rep* mutant constructed by CHEJANOVSKY and CARTER (1990). The mutant is defective for DNA replication but overexpresses both the p5 and p19 Rep proteins. One explanation of this phenotype is that a normal function of the Rep protein may be to attenuate its own synthesis; however, other explanations are also possible. The idea that the *rep* gene can autoregulate its own expression is attractive because it would provide a mechanism by which Rep protein synthesis (and the possibility of AAV DNA replication) could be reduced when the viral genome is in the proviral state.

The *rep* gene also has been found to be capable of repressing heterologous promoters. In both transient expression and DNA mediated transformation assays, selectable markers under the control of a variety of promoters are repressed when cotransfected with plasmids containing the wild type *rep* gene. These include the neomycin resistance gene under the control of the SV40 or mouse metallothionein promoters (LABOW et al. 1987), the neomycin resistance gene under AAV p40 control (MENDELSON et al. 1988b), the human immunodeficiency virus type 1 (HIV-1) genes driven by the HIV long terminal repeat (LTR) (ANTONI et al. 1991), and the herpes thymidine kinase gene (LABOW et al. 1987). In two of these cases, the HIV LTR and the SV40 promoter, repression appears to occur (at least in part) at the level of transcription (LABOW et al. 1987; ANTONI et al. 1991).

3.2 Helper Virus Functions

In the best studied case, adenovirus, five genetic regions are required for a fully permissive AAV infection. These are the E1A, E1B, E4, E2A, and VA regions. The E1A region is required for the *trans*-activation of the AAV p5 and p19 promoters, a 5–20-fold effect depending on the promoter (CHANG et al. 1989; TRATSCHIN et al. 1984b; LAUGHLIN et al. 1982). In the case of the p5 promoter, SHENK and his colleagues have recently reported evidence that at least two elements are involved in transcriptional control: a major late promoter sequence (UE) which is capable of binding the major late transcription factor

(MLTF) and the R1 sequence which is a novel E1A response element (CHANG and SHENK 1990). Together these two sequences appear to be essential for a full E1A induction of transcription from the p5 promoter. The mechanism of induction, as in the case adenovirus and cellular genes, is unknown.

The E1B-coded 55K transforming protein and the E4-coded 34K ORF-6 gene product are both required for the efficient accumulation of AAV mRNAs (CARTER et al. 1983; RICHARSON and WESTPHAL 1981, 1984; SAMULSKI and SHENK 1988; HUANG and HEARING 1989; LAUGHLIN et al. 1982, 1986; OSTROVE and BERNS 1980). It is believed that their role in AAV infection is similar to their role in adenovirus infection, i.e., they either stabilize the AAV mRNAs or they facilitate the transport of messages to the cytoplasm (SAMULSKI and SHENK 1988; HUANG and HEARING 1989; CUTT et al. 1987; HALBERT et al. 1985; SANDLER and KETNER 1989; SARNOW et al. 1984; BRIDGE and KETNER 1989). The E2A-coded adenovirus DNA binding protein (DBP) and the adenovirus VA RNAs have been shown to be required primarily for the efficient translation of the AAV p40 capsid mRNAs (WEST et al. 1987; JANIK et al. 1989; MCPHERSON et al. 1982; STRAUS et al. 1976b; JAY et al. 1981; MYERS et al. 1980; RICHARSON and WESTPHAL 1981, 1984). Again, a similar role has been defined for the E2A and VA genes during the expression of adenovirus genes (THIMMAPAYA et al. 1982; RICHARSON and WESTPHAL 1984). Mutations in the DNA binding protein do not seem to have much effect on AAV DNA replication and only a modest effect on the steady state level or stability of AAV transcripts (JAY et al. 1981; MYERS et al. 1980; QUINN and KITCHINGMAN 1986). However, the presence of the adenovirus DNA binding protein appears to be necessary for efficient splicing of AAV mRNA (WEST et al. 1987; JANIK et al. 1989). Further, in the absence of other adenovirus gene products it appears to be capable of inducing transcription from the AAV p5 promoter (CHANG and SHENK 1990). Finally, in addition to facilitating their translation, the presence of the VA RNA genes increases the steady state level of AAV transcripts (WEST et al. 1987; JANIK et al. 1989). The mechanism of this effect also is not understood.

3.3 Cellular Functions Required for AAV Replication

Although a fully permissive AAV infection requires the presence of a helper virus, no helper virus genes appear to be directly involved in AAV DNA replication. The role of the helper virus genes appears to be to maximize the synthesis of the AAV-coded gene products and possibly cellular genes that are required for AAV DNA replication. The best evidence for this is the fact that a small percentage of cells can become fully permissive for AAV DNA replication in the absence of helper virus if the cells are transformed with either a viral or a cellular oncogene and if the cells are also synchronized or treated with reagents that transiently arrest cellular DNA synthesis (e.g., hydroxyurea or UV treatment) (YALKINOGLU et al. 1988; BANTEL-SCHALL and ZUR HAUSEN 1988a, b; YACOBSON et al. 1987, 1989). In addition, mutations in the adenovirus DNA polymerase gene are normal for AAV DNA replication and DNA fragments which contain the adenovirus polymerase and adenovirus terminal protein coding regions are not required for AAV helper function in mixed transfection assays (STRAUS et al. 1976b; MYERS et al. 1980; JANIK et al. 1981). In vivo inhibitor studies suggest that an aphidocolin-sensitive cellular DNA polymerase is required for AAV DNA synthesis (MCLAUGHLIN and MUZYCZKA, in preparation), presumably polymerase α or δ (IKEGAMI et al. 1978; LEE et al. 1985). In vitro inhibitor studies of the terminal resolution reaction suggest that polymerase δ is required for this step (SNYDER et al. 1990b). Beyond this, however, nothing is known about the cellular gene products required for a productive AAV viral infection.

4 AAV Transduction Vectors

The use of AAV as a viral transduction vector was first demonstrated by HERMONAT and MUZYCZKA (1984). They deleted the AAV capsid gene between map positions 52 and 91 to make the vector *dl*52–91 (Fig. 5), and in place of the capsid gene they inserted the bacterial neomycin resistance gene under the control of the SV40 early promoter (Fig. 5, *dl*52–91/neo). A *dl*52–91/neo virus stock was obtained by transfecting the recombinant plasmid into human cells that had been infected with adenovirus (Fig. 6). To supply the missing capsid proteins it was necessary to cotransfect the cells with a second plasmid which contained a wild type *cap* gene but was itself defective for packaging (Fig. 6). This approach generated *dl*52–91/neo virus stocks that contained up to 10^6 infectious units/ml (HERMONAT and MUZYCZKA 1984). To inactivate the contaminating adenovirus, the recombinant stock was heated at 55–60° C for 2 h. Using this virus stock, HERMONAT and MUZYCZKA readily isolated Geneticin (G418) resistant cells when they infected established human cell lines. The transduction frequency of these stocks was approximately the same as the integration frequency for wild type virus, 0.5%–5.0% (LAUGHLIN et al. 1986; HANDA et al. 1977). TRATSCHIN et al. (1985) and LEBKOWSKI et al. (1988) subsequently constructed and tested similar vectors and obtained approximately the same transduction frequencies. It is worth noting that in their work, TRATSCHIN et al. (1985) placed the neomycin resistance gene under the control of the AAV p40 promoter and demonstrated that this promoter had a sufficiently high level of constitutive expression for Geneticin (G418) selection. Typically, the recombinant genome was found integrated into a host chromosome and was stable for up to 100 passages. However, if the proviral cells were superinfected with adenovirus, the recombinant provirus could be rescued and amplified (HERMONAT and MUZYCZKA 1984; TRATSCHIN et al. 1985; MCLAUGHLIN et al. 1988; SAMULSKI et al. 1989).

To determine the minimum AAV sequences required for transduction and to increase the available room for insertion of foreign DNA, MCLAUGHLIN et

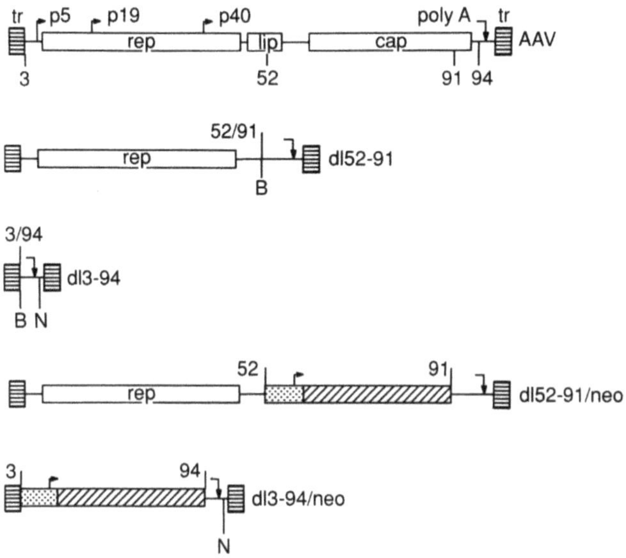

Fig. 5. Structure of AAV Recombinant Clones. A diagram of the wild type AAV genome and several AAV vectors and recombinants discussed in the text. The *horizontally hatched* boxes at the ends of each genome are the AAV terminal repeats (*tr*). The AAV polyadenylation site (*poly A*) is indicated by a *bent downward arrow* at map unit 95 (4425 bp). Locations of the three AAV promoters are indicated by *arrows* at map units 5 (255 bp), 19 (843 bp), and 40 (1822 bp). *Vertical lines* represent the site of the deletion in each vector. *dl*52–91 has a deletion from map unit 52 (2416 bp) to map unit 91 (4264 bp); *dl*3–94 contains a deletion from map unit 3 (144 bp) to 94 (4397 bp). The two AAV recombinant clones contain the SV40 enhancer (*stippled box*) and the neomycin phosphotransferase gene (*diagonally hatched box*) inserted in each vector at the site of the delection. The unique restriction sites in each vector, *Bgl*II (*B*) and *Nco*I (*N*), are also indicated. (Adapted from McLaughlin et al. 1988)

al. (1988) constructed the vector *dl*3–94 (Fig. 5). This vector contained only the left and right AAV terminal repeats and 139 bp of nonrepeated AAV sequences adjacent to the right terminal repeat. The nonrepeated sequences were retained because they contained the AAV polyadenylation signal and the vector also contained two cloning sites, one on either side of the polyadenylation signal (Fig. 5; McLaughlin et al. 1988). Samulski et al. (1987, 1989) constructed a similar vector, p*sub*201, in which only 45 bp of the nonrepeated sequences adjacent to the right terminal repeat were retained. Both vectors proved to be capable of efficiently transducing foreign DNA into human cells (McLaughlin et al. 1988; Samulski et al. 1989). This work proved that the terminal repeats (and possibly the sequences immediately adjacent to the right terminal repeat) were the only sequences required for AAV DNA replication, packaging of AAV viral DNA, integration of AAV proviruses, and rescue of integrated genomes. In addition, the vectors in which the *rep* gene had been removed were found to have up to an 80-fold higher transduction frequency than comparable *rep*$^+$ vectors (McLaughlin et al. 1988). Typical *rep*$^-$ transduction frequencies were

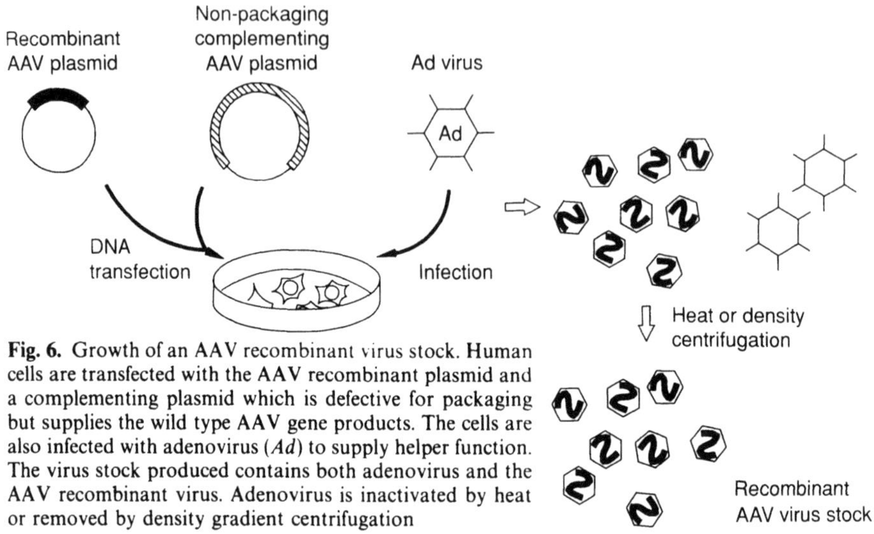

Fig. 6. Growth of an AAV recombinant virus stock. Human cells are transfected with the AAV recombinant plasmid and a complementing plasmid which is defective for packaging but supplies the wild type AAV gene products. The cells are also infected with adenovirus (*Ad*) to supply helper function. The virus stock produced contains both adenovirus and the AAV recombinant virus. Adenovirus is inactivated by heat or removed by density gradient centrifugation

50%–80% (McLaughlin et al. 1988; Samulski et al. 1989). The high transduction frequency of *rep*$^-$ vectors demonstrated that expression of the *rep* gene was not required for AAV integration. It also suggested that the limited level of *rep* gene expression that occurs in the absence of adenovirus either inhibits AAV integration or inhibits the expression of genes cloned into an AAV vector. The latter possibility was consistent with the evidence discussed earlier that the presence of the *rep* gene often resulted in the repression of heterologous gene expression during DNA-mediated transformation or transient expression assays (Labow et al. 1987; Mendelson et al. 1988b; Antoni et al. 1991). Vectors similar to *dl*3–94 have also been used by Lebkowski et al. (1988) and Vincent et al. (1990).

4.1 Size Constraints on AAV Vectors

Recombinant genomes that are between 50% and 110% of wild type AAV size can be efficiently packaged (Hermonat, McLaughlin and Muzyczka, unpublished; Samulski, personal communication; De La Maza and Carter 1980). This means that a vector such as *dl*3–94 can accommodate an insertion of 4.7 kb in length. Packaging of inserts of up to 5.2 kb (120% of wild type size) can also be detected but this occurs very inefficiently. Not surprisingly, recombinant genomes that are 50% of wild type AAV size or less can be packaged as multimeric genomic repeats (McLaughlin and Muzyczka, unpublished). In one study virus particles containing the *dl*3–94/neo genome (50% of wild type size) were found to consist predominantly of particles that had the density of AAV particles containing a wild type complement of DNA. When the DNA in these particles was examined it was found to consist of three kinds of molecules (McLaughlin and Muzyczka, unpublished). The predominant DNA molecule was an inverted dimer of the *dl*3–94/neo genome. Such molecules occur as a natural consequence of the AAV DNA replication mechanism and have been described in studies of spontaneously occurring deletions or AAV defective interfering (DI) genomes (de La Maza and Carter 1980a). In addition, some of the particles contained a tandem dimer of *dl*3–94/neo DNA which presumably formed by recombination during replication of the recombinant molecule. Finally, monomer length *dl*3–94/neo DNA was also found, suggesting that in some cases two different monomer length *dl*3–94/neo molecules were packaged into the same virus particle.

Although there is a restriction on the size of the DNA that can be packaged there seems to be no obvious restriction on the size of the recombinant AAV genome that can be rescued and replicated. In one study AAV recombinant DNA up to 24 kb in length (500% wild type size) was rescued from recombinant plasmids and replicated at approximately 10% of the wild type AAV DNA level (Wright and Muzyczka, unpublished). This means that in the presence of adenovirus helper functions the infectious AAV plasmid can be used for the transient expression of a foreign DNA insert of virtually any size (see for example, Wondisford et al. 1988). Further, rescue and replication would presumably occur even if the recombinant plasmid DNA was first integrated into a mammalian chromosome. AAV vectors, therefore, might be used in conjunction with other DNA transfer methods (i.e., microinjection or calcium phosphate transfection) to isolate cell lines that contain large integrated recombinant AAV genomes that are capable of rescue and expression by superinfection of the cell with adenovirus. This procedure has essentially been demonstrated by Mendelson et al. (1988a).

4.2 Poison Sequences

Several cases have been found in which a foreign DNA sequence inserted into an AAV vector has inhibited AAV DNA replication. In one case the insertion of a 3.4 kb fragment containing the herpes virus thymidine kinase gene reduced replication of the recombinant vector approximately 50-fold (Hermonat and Muzyczka, unpublished). In another case the insertion of a 15 kb fragment from the bacterial plasmid ColIb produced a recombinant in which no AAV DNA replication could be detected, whereas insertion of other ColIb fragments ranging in size from 5–20 kb had no effect on DNA replication (Wright and Muzyczka, unpublished). Finally, attempts to insert the adenovirus E4 region into the vector *dl*52–91 also failed to produce a recombinant that was capable of DNA replication (McLaughlin et al., unpublished). In the case of the E4 recombinant it was clear that the size of the recombinant was not an issue because a *dl*3–94/E4 genome which contained only the E4 open reading frames 5–7 and was approximately 90% of wild type size was defective for DNA replication. In addition, expression of the E4 sequences had no effect on DNA replication. Insertion of the E4 sequences downstream of the AAV p40 promoter in an orientation which did not allow expression of the E4 genes inhibited DNA replication as much as insertion of the E4 sequences in the opposite orientation. Furthermore, expression of the adenovirus E4 region from cotransfected plasmids did not inhibit wild type AAV DNA replication. Taken together, these experiments suggest that some sequences are incompatible with AAV DNA replication and inhibit replication in *cis*. The nature of these inhibitory sequences is not known.

4.3 Host Range

Relatively little is known about the natural course of AAV infections. Most of the adult human population (85%) is positive for serum antibodies to AAV capsid proteins and so, presumably, they have undergone an AAV infection, but no pathology or disease has been directly associated with AAV. Seroconversion usually occurs during childhood (BLACKLOW et al. 1967, 1968a, b, 1971; HOGGAN 1970; PARKS et al. 1970). AAV was originally identified as a contaminant of adenovirus preparations (ATCHISON et al. 1965; MELNICK et al. 1965; HOGGAN et al. 1966) and, thus far, has been isolated only from individuals that are undergoing the symptoms of an adenovirus infection (BLACKLOW et al. 1968a, b). Presumably, the natural route of AAV infections is the same as that of adenovirus, the intestinal or respiratory tracts. Although AAV inhibits the production of adenovirus in mixed infection (CASTO et al. 1967a, b), no clear correlation has been found between the severity of an adenovirus infection and the presence of a concurrent AAV infection (BLACKLOW et al. 1968a, 1971; LIPPS and MAYOR 1980, 1982). However, several studies suggest that AAV infection may protect individuals from virally induced oncogenesis (CASTO and GOODHEART 1972; CUCKOR et al. 1975; DE LA MAZA and CARTER 1981; MAYOR et al. 1973, 1976; OSTROVE et al. 1981; HERMONAT 1989).

Much of what is known about the host range of AAV comes from work with established tissue culture cells. It is generally believed that human AAV-2 can multiply productively in cells from any mammalian species provided an appropriate helper virus is present. For example, human AAV will replicate in canine cells infected with canine adenovirus (CASTO et al. 1967a). Further, whereas both human adenovirus and AAV are restricted in monkey cells for the synthesis of their respective capsid proteins, the restriction for both viruses can be overcome by coinfection with SV40 (BULLER et al. 1979; CHEJANOVSKY and CARTER 1989a). These observations suggest that the host range of the helper virus is the primary determinant for AAV host range during a productive viral infection. In contrast to productive infections, there may be a host range restriction for AAV integration. In one study comparing human, monkey, Chinese hamster, mink, and mouse cells, the transduction frequency of *dl*3–94/neo was 80%, 30%, 21%, 4%, and 0.4%, respectively (MCLAUGHLIN and MUZYCZKA, unpublished). However, in another study, the transduction of murine hematopoietic cells occurred at a frequency similar to that seen in human cells for the same recombinant virus (LAFACE et al. 1988). Finally, only a limited number of human cells have been examined for their ability to acquire an AAV provirus but all of them could be successfully transduced. They include several established transformed lines (HeLa, KB, D6, 293), a normal lymphoblastoid line (NC37), human T cell lines (CEM, H9), human colon cancer lines (HT29, LIM, CaCo), several leukemia lines (K562, KG1a, HEL, HL60, U937), and primary human liver hepatocytes (LEBKOWSKI et al. 1988; TRATSCHIN et al. 1985; MCLAUGHLIN et al. 1988; LAUGHLIN et al. 1986; SAMULSKI et al. 1989; SAMULSKI, personal communication).

4.4 Growth of Recombinant Viral Stocks

One of the difficulties in using AAV transduction vectors has been the awkward procedure required for growing a recombinant virus stock. The growth of a recombinant stock requires the presence of both AAV and helper virus genes and the recombinant viral stock that is produced contains both adenovirus and often wild type AAV virus particles as contaminants (Fig. 6). The use of a wild type AAV infectious plasmid to supply the AAV *rep* and *cap* gene products in *trans* produces stocks with unacceptably high levels of wild type AAV virus, approximately 10:1 wild type to recombinant (TRATSCHIN et al. 1985; LEBKOWSKI et al. 1988; VINCENT et al. 1990; HERMONAT and MUZYCZKA, unpublished). For reasons that are not clear there is a strong bias toward amplification of the wild type virus. The same is true when a recombinant viral stock is amplified by complementation with wild type AAV virus.

Several strategies have been tried to reduce the level of wild type AAV virus in recombinant stocks. HERMONAT and MUZYCZKA (1984) inserted a 2.5 kb fragment of lambda bacteriophage DNA into a nonessential region of the wild type AAV plasmid to produce a recombinant genome, *ins*96λ/M, which could replicate and supply all of the AAV gene products in *trans* but was itself too large to be packaged. Recombinant titers of 10^5–10^6 could be obtained but the stocks were contaminated with wild type virus at the level of 5%–10% (HERMONAT and MUZYCZKA 1984; MCLAUGHLIN et al. 1988). The contaminating wild type AAV virus was apparently the result of recombination between the complementing helper palsmid, *ins*96λ/M, and the AAV vector sequences. In addition, the bacteriophage sequences could spontaneously delete during the course of DNA replication (E. WINOCOUR, personal communication). Several laboratories have tried to use complementing AAV plasmids which are missing the terminal repeats and, therefore, are incapable of being packaged (TRATSCHIN et al. 1985; LEBKOWSKI et al. 1988; HERMONAT and MUZYCZKA, unpublished). This approach also produces significant levels of wild type contamination (1%–50%). In addition, the titers of the recombinant stocks are significantly lower (10^2–10^3), presumably because the complementing AAV genome cannot replicate and, therefore, produces lower levels of the AAV *rep* and capsid gene products (Samulski et al. 1989).

To eliminate the wild type contamination in recombinant stocks Samulski et al. (1989) constructed a complementing plasmid (pAAV/Ad) in which there were no homologous sequences between the recombinant genome and the complementing plasmid. This plasmid consisted of the AAV coding sequences flanked by the adenovirus 5 terminal repeats. The adenovirus termini apparently allowed the complementing AAV plasmid to undergo limited amplification after transfection into adenovirus-infected cells by the mechanism normally used for adenovirus DNA replication. The pAAV/Ad complementing plasmid produced recombinant virus titers of 10^4–10^5 with no detectable wild type contamination (Samulski et al. 1989).

Vincent et al. (1990) isolated several HeLa cell lines which contained integrated copies of the AAV genome that were missing the terminal repeats. The absence of the terminal repeats prevented the rescue and packaging of the integrated AAV sequences when the cells were superinfected with adenovirus. One of the lines (HA25a) was capable of generating recombinant stocks with titers of 10^3–10^4. Once again it was necessary to eliminate overlap between the vector sequences and the integrated AAV genome to prevent wild type contamination in the recombinant viral stocks. The low virus titers produced were apparently due to the low copy number of the wild type AAV genes in the packaging cell line. Mendelson et al. (1988a) also isolated several cell lines which constitutively expressed the AAV Rep proteins, suggesting that it might be possible to construct AAV packaging lines.

Chejanovsky and Carter (1989a) have reported the isolation of an amber mutant (pNTC3) in the AAV *rep* gene. The mutation could be efficiently suppressed by growing it on a monkey cell line containing an inducible human serine tRNA amber suppressor. The virus titers obtained were 10^7–10^8 (approximately 10% of the wild type titers obtained with the same monkey lines) and the reversion frequency of the amber mutation was less than 10^{-5}. Conceivably, the conditional lethal nonsense mutant could be used to complement the growth of AAV recombinant stocks or to construct an AAV packaging cell line.

The possibility of isolating a cell line which provides all of the adenovirus helper functions is more problematic. High expression of several of the adenovirus gene products required for AAV growth (particularly the adenovirus E4 region) is believed to be lethal to mammalian cells. However, it should be possible to use one or a combination of several adenovirus temperature-sensitive mutations (particularly those in the adenovirus DNA polymerase and in the adenovirus structural genes) to produce AAV vector stocks that are essentially free of adenovirus contamination. Thus far, two approaches have been reported for removing adenovirus from AAV stocks, heat inactivation and CsCl density centrifugation. The latter method is also an effective way of concentrating AAV virus stocks. Both methods sometimes leave detectable levels of adenovirus and it is often necessary to include anti-adenovirus neutralizing antibody in the cell media during infections with AAV vectors.

Finally, it should be mentioned that the AAV terminal repeat sequence is often unstable in plasmids propagated in standard *Rec*A prokaryotic hosts such as HB101 (BOYER 1969). The problem is particularly evident with vectors in which two terminal repeats are in close proximity, for example *dl*3–94. This problem can be solved by propagating AAV vectors in hosts such as JC8111 (BOISSY and ASTELL 1985) which is defective in several *E. coli* recombination pathways.

4.5 Expression of Foreign Genes in AAV Vectors

One of the potentially attractive features of rep^- AAV vectors is that, in the absence of the *rep* gene and adenovirus, the terminal repeats are expected to have no effect on the expression of foreign genes under the control of inducible promoters. Thus far, only a few inducible eukaryotic transcriptional control elements have been used to drive the expression of foreign genes in AAV vectors. WALSH et al. (1991) inserted the $^{A}\gamma$ globin gene into a rep^- AAV vector under the control of its own erythroid-specific locus control region (LCR site II). When the AAV-globin recombinant was transduced into erythroleukemia cells (K562), the basal and hemin-induced levels of expression from the transduced gene were equivalent to that of the endogenous native globin gene.

In a more extreme case, SRIVASTAVA et al. (1989) have inserted all of the coding and control regions of the B19 viral genome into an AAV rep^- vector. B19 is an autonomous human parvovirus whose growth is restricted to erythroid progenitor cells. The tissue restriction is apparently due to the presence of erythroid-specific transcriptional signals which control the expression of the B19 Rep protein. The B19 protein is homologous to the AAV Rep protein and, if it is expressed, B19 Rep will bind to the AAV terminal repeat of the hybrid AAV-B19 genome and promote its replication. When SRIVASTAVA et al. (1989) compared the ability of the hybrid AAV-B19 virus to replicate in human KB and human bone marrow cells, they found that the B19 Rep protein was expressed only in bone marrow cells. In fact, by every criteria examined, the AAV-B19 virus behaved like wild type B19 in bone marrow cells and like a rep^- AAV vector in KB cells. This suggests that the presence of the AAV terminal repeats does not alter the normal tissue specificity of the B19 control elements. In addition, the hybrid AAV-B19 virus may have some specialized advantages as a vector for bone marrow cells.

In addition to inducible control elements several constitutive eukaryotic promoters have been used successfully to drive the expression of foreign genes in rep^- AAV vectors. These include the SV40 early promoter (MCLAUGHLIN et al. 1988; SAMULSKI et al. 1989), the cytomegalovirus immediate early 1 (CMV IE-1) promoter (VINCENT et al. 1990), and the murine sarcoma virus LTR (LEBKOWSKI et al. 1988).

In contrast to rep^- vectors, the behavior of foreign control elements in rep^+ vectors is less predictable. As already mentioned, several laboratories have reported that the AAV *rep* gene product can inhibit the expression of foreign promoters in *trans* (ANTONI et al. 1991; LABOW et al. 1987; MENDELSON et al. 1988b). In addition, two reports suggest that in the presence of Rep protein, the terminal repeat can function as a control element which can induce transcription from either the AAV promoters (MCCARTY et al. 1991) or a foreign promoter (BEATON et al. 1989). However, other foreign promoters, such as the adenovirus E4 region, seem to be unaffected when they are inserted into a rep^+ vector (MCLAUGHLIN and MUZYCZKA, unpublished).

Finally, Carter and his colleagues have investigated the use of the AAV p40 promoter for transient gene expression (TRATSCHIN et al. 1984b, 1986; TREMPE and CARTER 1988a; WEST et al. 1987; WONDISFORD et al. 1988). In one study they compared the expression of the human thyrotropin β-subunit gene under the control of the AAV p40 promoter and the SV40 early promoter. Expression from the p40 promoter in 293 cells was found to be 10–100-fold higher than expression from the SV40 early promoter in monkey cells (WONDISFORD et al. 1988).

5 AAV Integration

The mechanism for AAV integration is not known and, thus far, no plausible model for integration has been proposed. The available evidence, however, suggests that the mechanism is different from that of any other known DNA element that recombines with host chromosomes. First, even though the integration frequencies of AAV genomes are among the highest reported for integrating DNA elements, no AAV gene expression is required to promote integration. As mentioned earlier, only the AAV terminal repeats appear to be essential for integration (SAMULSKI et al. 1989; MCLAUGHLIN et al. 1988). (It should be noted however that no one has yet eliminated the possibility that AAV virus particles contain an enzymatic activity that promotes integration.) Second, most AAV proviruses consist of at least one intact genome which can be rescued (LAUGHLIN et al. 1986; CHEUNG et al. 1980; MCLAUGHLIN et al. 1988). Third, a limited amplification of the AAV genome appears to be a requirement for integration (LAUGHLIN et al. 1986; MCLAUGHLIN et al. 1988; SAMULSKI et al. 1989). Finally, there is recent evidence that AAV integration is targeted to a specific chromosome (KOTIN et al. 1990).

The first indication that AAV could persist in cells as a latent infection was the observation by HOGGAN (1970) that AAV virus could be rescued from 20% of primary African green monkey and 2% of human embryonic kidney cells. Subsequently, HOGGAN et al. (1972) established a latent AAV infection experimentally in an established human tissue culture line and BERNS and his colleagues (CHEUNG et al. 1980; BERNS et al. 1975) demonstrated that AAV existed as an integrated provirus in latently infected cells. Much of what is known about AAV integration has come from genomic hybridization studies of some 80 independently isolated proviral lines (CHEUNG et al. 1980; TRATSCHIN et al. 1985; LAUGHLIN et al. 1986; McLAUGHLIN et al. 1988; SAMULSKI et al. 1989). Thus far, relatively little DNA sequence information has been reported (KOTIN and BERNS 1989; GROSSMAN et al. 1985) and a complete AAV provirus has not yet been isolated.

5.1 Proviral Structure

The predominant AAV proviral structure consists of two to four tandem copies of the AAV genome integrated in a single chromosome location (LAUGHLIN et al. 1986; McLAUGHLIN et al. 1988). The same frequency of concatemer formation is seen regardless of the multiplicity of infection. Further, even at high multiplicities of infection, only a limited number of separate chromosome integration events occurs. Tandem repeated copies of the AAV provirus can contain one or two copies of the terminal repeat sequence between two adjacent AAV genomes but a mixture of one and two copies has not been observed in the same tandem array (McLAUGHLIN et al. 1988). In addition tandem arrays of AAV genomes that contain identical internal deletions have been found (McLAUGHLIN et al. 1988; LAUGHLIN et al. 1986). These are apparently due to the amplification of a rare mutant contaminating the AAV virus stock and are not likely to have occurred by recombination of several input viral genomes. All of these observations suggest that the tandem arrays are due to the replication of a single AAV genome rather than the end to end joining of multiple input AAV molecules (McLAUGHLIN et al. 1988; LAUGHLIN et al. 1986). Further, the replication events associated with integration probably occur by a different mechanism than that used during normal AAV DNA replication. Normal AAV replication requires the presence of the Rep protein and produces concatemers which are head to head or tail to tail (Fig. 4).

In addition to tandem concatemers three other kinds of proviral structures have also been found (McLAUGHLIN et al. 1988; LAUGHLIN et al. 1986). A small number of cell lines have been isolated which apparently contain a single integrated copy of the AAV genome and some of these are also capable of rescue. However, because of the limitations of genomic hybridization it is not certain whether these proviruses contain a partial duplication of the AAV genome. In addition, several cell lines have been isolated in which both the AAV sequences and a segment of flanking cellular DNA have been amplified to form a tandem concatemer (McLAUGHLIN et al. 1988). Finally, in a number of cases cell lines that contain an integrated provirus and no free AAV DNA were found after repeated passage to contain unit length linear AAV DNA in an episomal form (CHEUNG et al. 1980; McLAUGHLIN et al. 1988). It is not clear whether the extrachromosomal copies represent integrated copies that had excised in some fraction of the cell population or whether the integrated copy had been selectively replicated to form an episomal copy.

One reason why concatemers may be a necessary feature of AAV integration is that recombination between AAV and cellular sequences may not be precise. Although most cellular-AAV junctions appear to be near an AAV end, in some cases integration has clearly occurred randomly with respect to AAV sequences (McLaughlin et al. 1988). This has been confirmed by Kotin and Berns (1989) who have sequenced two AAV-cellular junctions. In both cases recombination with cellular DNA had occurred within the AAV terminal repeat but a substantial portion of the terminal sequence had been deleted. Grossman et al. (1984, 1985) have also demonstrated that recombination at the terminal repeat may be imprecise. Using a system in which they could study recombination between AAV DNA and the SV40 chromosome, they demonstrated that crossover generally occurred within the terminal repeat but left a terminal sequence that had been partially deleted (Grossman et al. 1985). Recently, several AAV-AAV junctions that are present in tandem proviral arrays have been cloned and sequenced. In contrast to the cellular-AAV junctions, the terminal repeats between two AAV copies of a tandem proviral array contain an intact AAV terminal sequence (Zolotokhin and Muzyczka, unpublished). It is possible, therefore, that the formation of a concatemer is a mechanism for insuring the integrity of at least one copy of the AAV genome during integration.

The work of Grossman et al. (1984) also demonstrated another feature of AAV integration. When they examined the SV40-AAV recombinants formed after cotransfection with AAV and SV40 DNA, the AAV sequences were random. In contrast, when the SV40 and AAV DNA was delivered to cells by viral infection, the AAV sequences present in the recombinants consisted predominantly of AAV terminal sequences. This suggested that delivery of AAV DNA to cells by virus infection was necessary for some aspect of AAV integration that involved the terminal repeats. Either there was a requirement for a single-stranded AAV DNA molecule that is normally present in particles or some component of the AAV capsid had a function in integration.

5.2 Site-Specific Integration

The analysis of AAV proviral clones by genomic hybridization had suggested that AAV integration was random with respect to cellular DNA sequences (McLaughlin et al. 1988; Laughlin et al. 1986). This appeared to be true regardless of whether selection had been applied in isolating the line. However, recent results presented by Berns and his colleagues (Kotin et al. 1990) suggest that there is a clear specificity to AAV integration. Using flanking DNA probes obtained from two clones containing AAV-cellular junctions, they discovered that most AAV proviruses (68% of those tested) contained the same flanking cellular DNA sequences. They further located the cellular target sequence by somatic cell hybrid mapping to a 7-kb region of human chromosome 19. The specificity of integration may have been missed in earlier studies because a substantial amount of rearrangement occurs in the cellular target sequence during AAV integration; this apparently is the case in one cell line (KOTIN and BERNS 1989). In addition, there may be multiple target sequences for integration within the 7-kb region of chromosome 19.

The work of BERNS and his colleagues has been confirmed by SAMULSKI et al. (1991). SAMULSKI and his colleagues have independently isolated a flanking cellular DNA probe and demonstrated that it has the same chromosome 19 sequence contained in the flanking DNA isolated by KOTIN and BERNS (1989). Using this probe SAMULSKI et al. (1991) extended the analysis to a number of additional proviral lines and showed that, indeed, all of the *rep*$^+$ proviruses thus far examined have integrated into a chromosome 19 sequence. In addition, both groups (SAMULSKI et al. 1991; KOTIN et al. 1991) have used in situ chromosome hybridization to directly map the proviral integration site in several cell lines to q13.4-ter of chromosome 19.

SAMULSKI and his colleagues may also have found a partial explanation for why some proviruses are not associated with the chromosome 19 target sequence (SAMULSKI, personal communication). Using genomic blotting they found that although all *rep*$^+$ viruses were associated with the chromosome 19 sequence, most of the *rep*$^-$ proviruses were not. In one set of proviral lines that were generated with *rep*$^-$ virus stocks containing no wild type AAV contamination, eight out eight proviruses were found not to be associated with the chromosome 19 sequence. This suggested that the integration site was different when the Rep protein was not expressed during an AAV infection. Thus far, two *rep*$^-$ proviruses have been mapped by in situ hybridization to chromosomes 22 and 12.

Taken together the current evidence suggests the following. First, after infection with AAV virus, the AAV termini are efficiently recognized by cellular enzymes as an integration signal and also as a signal for limited amplification. Second, in the presence of the *rep* gene, integration appears to be specific for a cellular sequence in chromosome 19. Finally, in the absence of Rep protein, when integration is most efficient, there either are additional cellular target sequences for integration or integration is random.

5.3 Rescue of Proviral Sequences

No convincing model for AAV rescue has been proposed. Most of the speculation has focused on the possibility that the AAV Rep protein can initiate excision by making the same cuts that it makes during terminal resolution and several models have been suggested to explain how this might occur (SAMULSKI et al. 1983; SENEPATHY et al. 1984). In addition, GOTTLIEB and MUZYCZKA (1988, 1990a, b) have isolated a cellular enzyme, endo R, that is capable of excising AAV sequences from prokaryotic plasmids at low frequencies by producing double-stranded cuts within the first 15 bp of the AAV terminal repeat. The major activity of endo R is to produce double-stranded cuts within any sequence which is rich in G:C bp. These sequences have been shown to adopt triple-stranded DNA structures that involve non-canonical G:G base pairing (GOTTLIEB and MUZYCZKA 1990b and references therein). It is possible that G-rich sequences are hotspots for mammalian recombination and that endo R initiates cellular recombination at these sequences. In this respect it is worth noting that the proviral cellular integration site cloned by KOTIN and BERNS (1989) also contains several G-rich sequences that are excellent substrates for endo R. Nevertheless, there is no convincing evidence that either the AAV Rep protein or endo R is directly involved in the rescue of AAV DNA from mammalian chromosomes or from prokaryotic plasmids.

6 Gene Therapy

The *rep*⁻ AAV vectors have several features that make them attractive candidates for human gene therapy. First, the cloning capacity of 5 kb can accommodate a variety of cDNAs. Second, the frequency of transduction in human cells is high. So far, no human cell line or tissue has been shown to be resistant to AAV transduction. Third, no disease has been associated with AAV in either human or animal populations. In addition, because AAV vectors require two other viral genomes to be propagated (wild type AAV and adenovirus), there is an inherent limit to the natural spread of an AAV vector. Fourth, AAV proviruses appear to be stable. In addition, if a proviral cell is superinfected with adenovirus, a *rep*⁻ AAV provirus is incapable of DNA replication unless a wild type AAV genome is also supplied (MCLAUGHLIN et al. 1988). Individuals carrying an AAV provirus can also be protected from adenovirus infection by vaccination. Fifth, in the absence of the *rep* gene, the AAV terminal repeats appear to be transcriptionally neutral. Thus, AAV vectors may be useful when it is essential to have foreign genes under the control of their own enhancer and promoter elements. Finally, it is worth noting that there is no superinfection immunity for AAV vectors. A cell line can be transduced multiple times with several different AAV vectors (LEBKOWSKI et al. 1988; MCLAUGHLIN et al. 1988).

In spite of these features there are still several problems associated with AAV vectors that need to be resolved. First, the procedures for growing recombinant virus stocks continue to be awkward. Fortunately, this does not appear to be an insurmountable technical problem. Second, the mechanism for AAV integration is not known. Until this question is resolved it will be difficult to use AAV vectors for gene therapy with confidence. The fact that integration occurs at a limited number of chromosome sites (perhaps only one) is an attractive feature of AAV vectors, but it is still not clear whether this is true of all AAV vectors. Finally, relatively little is known about the effect of AAV infection or integration on primary cells. It is also not known whether primary cells must undergo a round of cell division for integration to occur. Although AAV infection appears to have no effect on the cell viability of established tissue culture lines, a recent report suggested that the growth of some primary cells may be inhibited by AAV infection (WINOCOUR et al. 1988).

Acknowledgements. The author thanks R.J. SAMULSKI for his suggestions on the manuscript. He also thanks R. KOTIN, R.J. SAMULSKI, B.J. CARTER, T. SHENK, A. SRIVASTAVA, and K.I BERNS for sharing the results of their experiments prior to publication. The authors's work was supported by grants from the NIH (RO1GM3572302) and the NCI (PO1CA2814607).

References

Antoni BA, Rabson AB, Miller IL, Trempe JP, Chejanovsky N, Cater BJ (1991) Adeno-associated virus Rep protein inhibits human immunodeficiency virus type 1 production in human cells. J Virol 65: 396–404

Ashktorab H, Srivastava A (1989) Identification of nuclear proteins that specifically interact with the adeno-associated virus 2 inverted terminal repeat hairpin DNA. J Virol 63: 3034–3039

Atchison RW, Casto BC, Hammond WMcD (1965) Adenovirus-associated defective virus particles. Science 149: 754–756

Bachmann PA, Hoggan MD, Kurstak E, Melnick JL, Pereira HG, Tattersall P, Vago C (1979) Parvoviridae: second report. Intervirology 11: 248–254

Bantel-Schall U, zur Hausen H (1988a) Adeno-associated viruses inhibit SV40 DNA amplification and replication of Herpes simplex virus in SV40-transformed hamster cells. Virology 164: 64–74

Bantel-Schall U, zur Hausen H (1988b) Dissociation of carcinogen-induced amplification and amplification of AAV DNA in a Chinese hamster cell line. Virology 166: 113–122

Beaton A, Palumbo P, Berns KI (1989) Expression from the adeno-associated virus p5 and p19 promoters is negatively regulated in *trans* by the *rep* protein. J Virol 63: 4450–4454

Becerra SP, Rose JA, Hardy M, Baroudy BM, Anderson CW (1985) Direct mapping of adeno-associated virus proteins B and C: a possible ACG initiation codon. Proc Natl Acad Sci USA 82: 7919–7923

Becerra SP, Koczot F, Fabisch P, Rose JA (1988) Synthesis of adeno-associated virus structural proteins requires both alternative mRNA splicing and alternative initiations from a single transcript. J Virol 62: 2745–2754

Berns KI (1990) Parvorviridae and their replication. In: Fields BN, Knipe DM (eds) Virology, vol 2, 2nd edn. Raven, New York, pp 1743–1763

Berns KI, Adler S (1972) Separation of two types of adeno-associated virus particles containing complementary polynucleotide chains. J Virol 5: 693–699

Berns KI, Bohenzky RA (1987) Adeno-associated viruses: an update. Adv Virus Res 32: 243–306

Berns KI, Hauswirth WW (1979) Adeno-associated viruses. Adv Virus Res 25: 407–449

Berns KI, Kelly TJ Jr (1974) Visualization of the inverted terminal repetition in adeno-associated virus DNA. J Mol Biol 82: 267–271

Berns KI, Rose JA (1970) Evidence for a single-stranded adeno-associated virus genome: isolation and separation of complementary single strands. J Virol 5: 693–699

Berns KI, Pinkerton TC, Thomas GF, Hoggan MD (1975) Detection of adeno-associated virus (AAV)-specific nucleotide sequences in DNA isolated from latently infected Detroit 6 cells. Virology 68: 556–560

Blacklow NR, Hoggan MD, Rowe WP (1967) Isolation of adenovirus-associated viruses from man. Proc Natl Acad Sci USA 58: 1410–1415

Blacklow NR, Hoggan MD, Kapikian AZ, Austin JB, Rowe WP (1968a) Epidemiology of adenovirus-associated virus infection in a nursery population. Am J Epidermiol 8: 368–378

Blacklow NR, Hoggan MD, Rowe WP (1968b) Serological evidence for human infection with adenovirus-associated viruses. JNCI 40: 319–327
Blacklow NR, Hoggan MD, Sereno MS, Brandt CD, Kim HW, Parrott RH, Chanock RM (1971) A seroepidemiologic study of adenovirus-associated virus infections in infants and children. Am J Epidemiol 94: 359–366
Bohenzky RA, LeFebvre RB, Berns KI (1988) Sequence and symmetry requirements within the palindromic sequences of the adeno-associated virus terminal repeat. Virology 166: 316–327
Boissy R, Astell CR (1985) *Escherichia coli recBC sbcB recF* hosts permits the deletion-resistant propagation of plasmid clones containing the 5′-terminal palindrome of minute virus of mice. Gene 35: 179–185
Boyer HW (1969) A complementation analysis of the restriction and modification of DNA in *Eschericia coli*. J Mol Biol 41: 459–479
Bridge E, Ketner G (1989) Redundant control of adenovirus late gene expression by early region 4. J Virol 63: 631–638
Buller RM, Straus SE, Rose JA (1979) Mechanism of host restriction of adenovirus-associated virus replication in African green monkey kidney cells. J Gen Virol 43: 663–672
Buller RM, Janik JE, Sebring ED, Rose JA (1981) Herpes simplex virus types 1 and 2 completely help adenovirus-associated virus replication. J Virol 40: 241–247
Carter BJ, Marcus-Sekura CJ, Laughlin CA, Ketner G (1983) Properties of an adenovirus type 2 mutant, Ad*dl*807, having a deletion near the right-hand genome terminus: failure to help AAV replication. Virology 126: 505–516
Cassinotti P, Weitz M, Tratschin JD (1989) Organization of the adeno-associated virus (AAV) capsid gene: mapping of a minor spliced mRNA coding for the virus capsid protein 1. Virology 167: 176–184
Casto BC, Goodheart CR (1972) Inhibition of adenovirus transformation in vitro by AAV-1. Proc Soc Exp Biol Med 140: 72–78
Casto BC, Atchison RW, Hammon WMcD (1967a) Studies on the relationship between adeno-associated virus type 1 (AAV-1) and adenoviruses. I. Replication of AAV-1 in certain cell cultures and its effect on helper adenoviruses. Virology 32: 52–59
Casto BC, Armstrong JA, Atchison RW, Hammon WMcD (1967b) Studies on the relationship between adeno-associated virus type 1 (AAV-1) and adenoviruses. II. Inhibition of adenovirus plaques by AAV; its nature and specificity. Virology 33: 452–458
Cavalier-Smith T (1974) Palindromic base sequences and replication of eukaryotic chromosome ends. Nature 250: 467–470
Chang L-S, Shenk T (1990) The adenovirus DNA binding protein stimulates the rate of transcription directed by adenovirus and adeno-associated virus promoters. J Virol 64: 2103–2109
Chang L-S, Shi Y, Shenk T (1989) Adeno-associated virus p5 promoter contains an adenovirus E1A inducible element and a binding site for the major late transcription factor. J Virol 63: 3479–3488
Chejanovsky N, Carter BJ (1989a) Replication of a human parvovirus nonsense mutant in mammalian cells containing an inducible amber suppressor. Virology 171: 239–247
Chejanovsky N, Carter BJ (1989b) Mutagenesis of an AUG codon in the adeno-associated virus rep gene: effects on viral DNA replication. Virology 173: 120–128
Chejanovsky N, Carter BJ (1990) Mutation of a consensus nucleotide binding site in the adeno-associated virus rep gene generates a dominant negative phenotype for DNA replication. J Virol 64: 1764–1770
Cheung AK, Hoggan MD, Hauswirth WW, Berns KI (1980) Integration of the adeno-associated virus genome into cellular DNA in latently infected human Detroit 6 cells. J Virol 33: 739–748
Cukor G, Blacklow NR, Kibrick S, Swan IC (1975) Effects of adeno-associated virus on cancer expression by herpesvirus-transformed hamster cells. JNCI 55: 957–959
Cutt JR, Shenk T, Hearing P (1987) Analysis of adenovirus early region 4-encoded polypeptides synthesized in productively infected cells. J Virol 61: 543–552
de la Maza LM, Carter BJ (1980a) Molecular structure of adeno-associated virus variant DNA. J Biol Chem 255: 3194–3203
de la Maza LM, Carter BJ (1980b) Heavy and light particles of adeno-associated virus. J Virol 33: 1129–1137
de la Maza LM, Carter BJ (1981) Inhibition of adenovirus oncogenicity in hamsters by adeno-associated virus DNA. JNCI 67: 1323–1326

Flint J, Shenk T (1989) Adenovirus E1A protein paradigm viral transactivator. Annu Rev Genet 23: 141–161
Gerry HW, Kelly TJ Jr, Berns KI (1973) Arrangement of nucleotide sequences in adeno-associated virus DNA. J Mol Biol 79: 207–225
Gottlieb J, Muzyczka N (1988) In vitro excision of adeno-associated virus DNA from recombinant plasmids: isolation of an enzyme fraction from HeLa cells that cleaves DNA at poly(G) sequences. Mol Cell Biol 6: 2513–2522
Gottlieb J, Muzyczka N (1990a) Purification and characterization of HeLa endonuclease R: a G-specific mammalian endonuclease. J Biol Chem 265: 10836–10841
Gottlieb J, Muzyczka N (1990b) Substrate specificity of HeLa endonuclease R: a G-specific mammalian endonuclease. J Biol Chem 265: 10842–10850
Green MR, Roeder RG (1980a) Transcripts of the adeno-associated virus genome: mapping of the major RNAs. J Virol 36: 79–92
Green MR, Roeder RG (1980b) Definition of a novel promoter for the major adenovirus-associated virus mRNA. Cell 1: 231–242
Green MR, Straus SE, Roeder RG (1980) Transcripts of the adenovirus-associated virus genome: multiple polyadenylated RNAs including a potential primary transcript. J Virol 35: 560–565
Grossman Z, Winocour E, Berns KI (1984) Recombination between simian virus 40 and adeno-associated virus: virion coinfection compared to DNA contransfection. Virology 134: 125–137
Grossman Z, Berns KI, Winocour E (1985) Structure of simian virus 40-adeno-associated virus recombinant genomes. J Virol 56: 457–465
Halbert DH, Cutt JR, Shenk T (1985) Adenovirus early region 4 encodes functions required for efficient DNA replication, late gene expression and host cell shutoff. J Virol 56: 250–257
Handa H, Shiroki K, Shimojo H (1977) Establishment and characterization of KB cell lines latently infected with adeno-associated virus type 1. Virology 82: 84–92
Hauswirth WW, Berns KI (1977) Origin and termination of adeno-associated virus DNA replication. Virology 78: 488–499
Hauswirth WW, Berns KI (1979) Adeno-associated virus DNA replication. Virology 93: 57–68
Hermonat PL (1989) The adeno-associated virus Rep78 gene inhibits cellular transformation by bovine papillomavirus. Virology 172: 253–261
Hermonat PL (1991) Inhibition of H-ras expression by the adeno-associated virus Rep78 transformation suppressor gene product. Cancer Research 51: 3373–3377
Hermonat PL, Muzyczka N (1984) Use of adeno-associated virus as a mammalian DNA cloning vector: transduction of neomycin resistance into mammalian tissue culture cells. Proc Natl Acad Sci USA 81: 6466–6470
Hermonat PL, Labow MA, Wright R, Berns KI, Muzyczka N (1984) Genetics of adeno-associated virus: isolation and preliminary characterization of adeno-associated virus type 2 mutants. J Virol 51: 329–333
Hoggan MD (1970) Adeno-associated viruses. Prog Med Virol 12: 211–239
Hoggan MD, Blacklow NR, Rowe WP (1966) Studies of small DNA viruses found in various adenovirus preparations: physical, biological, and immunological characteristics. Proc Natl Acad Sci USA 55: 1457–1471
Hoggan MD, Thomas GF, Thomas FB, Johnson FB (1972) Continuous carriage of adenovirus associated virus genome in cell culture in the absence of helper adenovirus. In: Proceedings of the fourth lepetite colloquium. Cocoyac, Mexico. North-Holland, Amsterdam, pp 243–249
Huang M-M, Hearing P (1989) Adenovirus early region 4 encodes two gene products with redundant effects in lytic infection. J Virol 63: 2605–2615
Ikegami S, Taguchi T, Ohashi M (1979) Aphidocolin prevents mitotic cell division by interfering with the activity of DNA polymerase-α. Nature 275: 458–459
Im D-S, Muzyczka N (1989) Factors that bind to the AAV terminal repeats. J Virol 63: 3095–3104
Im D-S, Muzyczka N (1990) The AAV origin binding protein Rep68 is an ATP-dependent site-specific endonuclease with DNA helicase activity. Cell 61: 447–457
Janik JE, Huston MM, Rose JA (1981) Locations of adenovirus genes required for the replication of adenovirus-associated virus. Proc Natl Acad Sci USA 78: 1925–1929
Janik JE, Huston MM, Rose JA (1984) Adeno-associated virus proteins: origin of the capsid components. J Virol 52: 591–597
Janik JE, Huston MM, Cho K, Rose JA (1989) Efficient synthesis of adeno-associated virus structural proteins requires both adenovirus DNA binding protein and VA I RNA. Virology 168: 320–329

Jay FT, Laughlin CA, Carter BJ (1981) Eukaryotic translational control: adeno-associated virus protein synthesis is affected by a mutation in the adenovirus DNA-binding protein. Proc Natl Acad Sci USA 78: 2927–2931
Johnson FB, Ozer HL, Hoggan MD (1971) Structural proteins of adenovirus-associated virus type 3. J Virol 8: 860–863
Johnson FB, Whitaker CW, Hoggan MD (1975) Structural polypeptides of adenovirus-associated virus top component. Virology 65: 196–203
Johnson FB, Thomson TA, Taylor PA, Vlazny DA (1977) Molecular similarities among the adenovirus-associated virus polypeptides and evidence for a precursor protein. Virology 82: 1–13
Khlief SN, Myers T, Carter BJ, Trempe JP (1991) Inhibition of cellular transformation by the adeno-associated virus rep gene. Virology 181: 738–741
Kirschtein RL, Smith KO, Peters EA (1968) Inhibition of adenovirus 12 oncogenicity by adeno-associated virus. Proc Soc Exp Biol Med 128: 3322–3330
Koczot FJ, Carter BJ, Garon CF, Rose JA (1973) Self-complementarity of terminal sequences within plus or minus strands of adenovirus-associated virus DNA. Proc Natl Acad Sci USA 70: 215–219
Kotin RM, Berns KI (1989) Organization of adeno-associated virus DNA in latently infected Detroit 6 cells. Virology 170: 460–467
Kotin RM, Siniscalco M, Samulski RJ, Zhu X, Hunter L, Laughlin CA, McLaughlin S, Muzyczka N, Rocchi M, Berns KI (1990) Site-specific integration by adeno-associated virus. Proc Natl Acad Sci USA 87: 2211–2215
Kotin RM, Menninger JC, Ward DC, Berns KI (1991) Mapping and direct visualization of a region-specific viral DNA integration site on chromosome 19q13-qter. Genomics 10: 831–834.
Labow MA, Hermonat PL, Berns KI (1986) Positive and negative autoregulation of the adeno-associated virus type 2 genome. J Virol 60: 251–258
Labow MA, Graf LH, Berns KI (1987) Adeno-associated virus gene expression inhibits cellular transformation by heterologous genes. Mol Cell Biol 7: 1320–1325
LaFace D, Hermonat PL, Wakeland E, Peck A (1988) Gene transfer into hematopoietic progenitor cells mediated by an adeno-associated virus vector. Virology 162: 483–486
Laughlin CA, Westphal H, Carter BJ (1979) Spliced adenovirus-associated virus RNA. Proc Natl Acad Sci USA 76: 5567–5571
Laughlin CA, Jones N, Carter BJ (1982) Effects of deletions in adenovirus region 1 genes upon replication of adeno-associated virus. J Virol 41: 868–876
Laughlin CA, Tratschin J-D, Coon H, Carter BJ (1983) Cloning of infectious adeno-associated virus genomes in bacterial plasmids. Gene 23: 65–73
Laughlin CA, Cardellichio CB, Coon HC (1986) Latent infection of KB cells with adeno-associated virus type 2. J Virol 60: 515–524
Lebkowski JS, McNally MM, Okarma TB, Lerch LB (1988) Adeno-associated virus: a vector system for efficient introduction of DNA into a variety of mammalian cell types. Mol Cell Biol 8: 3988–3996
Lee MYWT, Toomey NL, Wright GE (1985) Differential inhibition of human placental DNA polymerases α and δ by BuPdGTP and BuAdATP. Nucleic Acids Res 13: 8623–8630
LeFebvre RB, Riva S, Berns KI (1984) Conformation takes precedence over sequence in adeno-associated virus DNA replication. Mol Cell Biol 4: 1416–1419
Lipps BV, Mayor HD (1980) Transplacental infection with adeno-associated virus type 1 in mice. Intervirology 14: 118–123
Lipps BV, Mayor HD (1982) Defective parvoviruses acquired via the transplacental route protect mice against lethal adenovirus infection. Infect Immun 37: 200–204
Lusby E, Berns KI (1982) Mapping of the 5′ termini of two adeno-associated virus 2 RNAs in the left half of the genome. J Virol 41: 518–526
Lusby E, Fife KH, Berns KI (1980) Nucleotide sequence of the inverted terminal repetition in adeno-associated virus DNA. J Virol 34: 402–409
Lusby E, Bohenzky R, Berns KI (1981) Inverted terminal repetitions in adeno-associated virus DNA: independence of the orientation at either end of the genome. J Virol 37: 1083–1086
Mayor HD, Torikai K, Melnick J, Mandel M (1969) Plus and minus single-stranded DNA separately encapsidated in adeno-associated satellite virions. Science 166: 1280–1282
Mayor HD, Houlditch GS, Mumford DM (1973) Influence of adeno-associated virus on adenovirus-induced tumors in hamsters. Nature New Biol 241: 44–46
Mayor HD, Drake S, Stahmann J, Mumford DM (1976) Antibodies to adeno-associated satellite

virus and herpes simplex in sera from cancer patients and normal adults. Am J obstet Gynecol 126: 100–104
McCarty DM, Christensen M, Muzyczka N (1991) Sequences required for the coordinate induction of the AAV p19 and p40 promoters by the Rep protein. J Virol 65: 2936–2945
McLaughlin SK, Collis P, Hermonat PL, Muzyczka N (1988) Adeno-associated virus general transduction vectors: analysis of proviral structures. J Virol 62: 1963–1973
McPherson RA, Rose JA (1983) Structural proteins of adenovirus-associated virus: subspecies and their relatedness. J Virol 46: 523–529
McPherson RA, Ginsburg HS, Rose JA (1982) Adeno-associated virus helper activity of adenovirus DNA binding protein. J Virol 44: 666–673
McPherson RA, Rosenthal LJ, Rose JA (1985) Human cytomegalovirus completely helps adeno-associated virus replication. Virology 147: 217–222
Melnick JL, Mayor HD, Smith KO, Rapp F (1965) Association of 20 millimicron particles with adenoviruses. J Bacteriol 90: 271–274
Mendelson E, Trempe JP, Carter BJ (1986) Identification of the trans-active rep proteins of adeno-associated virus by antibodies to a synthetic oligopeptide. J Virol 60: 823–832
Mendelson E, Miller IL, Carter BJ (1988a) Expression and rescue of a nonselected marker from an integrated AAV vector. Virology 166: 154–165
Mendelson E, Smith MG, Miller IL, Carter BJ (1988b) Effect of a viral rep gene on transformation of cells by an adeno-associated virus vector. Virology 166: 612–615
Myers MW, Carter BJ (1980) Assembly of adeno-associated virus. Virology 102: 71–82
Myers MW, Laughlin CA, Jay FT, Carter BJ (1980) Adenovirus helper function for growth of adeno-associated virus: effect of temperature-sensitive mutations in adenovirus early gene region 2. J Virol 35: 65–75
Ostrove JM, Berns KI (1980) Adenovirus early region 1b gene function required for rescue of latent adeno-associated virus. Virology 104: 502–506
Ostrove JM, Duckworth DH, Berns KI (1981) Inhibition of adenovirus-transformed cell oncogenicity by adeno-associated virus. Virology 113: 521–533
Parks WP, Boucher DW, Melnick JL, Taber LH, Yow MD (1970) Seroepidemiological and ecological studies of the adenovirus-associated satellite viruses. Infect Immun 2: 716–722
Pattison JR (1990) Parvoviruses medical and biological aspects. In: Fields BN, Knipe DM (eds) Virology, vol 2, 2nd edn. Raven, New York, pp 1765–1784
Quinn CO, Kitchingman GR (1986) Functional analysis of the adenovirus type 5 DNA-Binding protein: site-directed mutants which are defective for adeno-associated virus helper activity. J Virol 60: 653–661
Redemann BE, Mendelson E, Carter BJ (1989) Adeno-associated virus Rep protein synthesis during productive infection. J Virol 63: 873–882
Richardson WD, Westphal WD (1981) A cascade of adenovirus early functions is required for expression of adeno-associated virus. Cell 27: 133–141
Richardson WD, Westphal WD (1984) Requirement for either early region 1a or early region 1b adenovirus gene products in the helper effect for adeno-associated virus. J Virol 51: 404–410
Rose JA, Koczot FJ (1972) Adenovirus-associated virus multiplication: VII. Helper requirement for viral deoxyribonucleic acid and ribonucleic acid synthesis. J Virol 10: 1–8
Rose JA, Berns KI, Hoggan MD, Koczot FJ (1969) Evidence for a single-stranded adenovirus-associated virus genome: Formation of a DNA density hybrid on release of viral DNA. Proc Natl Acad Sci USA 64: 863–869
Rose JA, Maizel JK, Shatkin AJ (1971) Structural proteins of adenovirus-associated viruses. J Virol 8: 766–770
Samulski RJ, Shenk T (1988) Adenovirus E1B 55-M_r polypeptide facilitates timely cytoplasmic accumulation of adeno-associated virus mRNAs. J Virol 62: 206–210
Samulski RJ, Berns KI, Tan M, Muzyczka N (1982) Cloning of adeno-associated virus into pBR322: rescue of intact virus from the recombinant plasmid in human cells. Proc Natl Acad Sci USA 79: 2077–2081
Samulski RJ, Srivastava A, Berns KI, Muzyczka N (1983) Rescue of adeno-associated virus from recombinant plasmids: gene correction within the terminal repeats of AAV. Cell 33: 135–143
Samulski RJ, Chang L-S, Shenk T (1987) A recombinant plasmid from which an infectious adeno-associated virus genome can be excised in vitro and its use to study viral replication. J Virol 61: 3096–3101
Samulski RJ, Chang L-S, Shenk T (1989) Helper-free stocks of recombinant adeno-associated

viruses: normal integration does not require viral gene expression. J Virol 63: 3822–3828
Samulski RJ, Zhu X, Xia X, Brook JD, Housman DE, Epstein N, Hunter LA (1991) Targeted integration of adeno-associated virus (AAV) into human chromosome 19. EMBO Journal, in press
Sandler AB, Ketner G (1989) Adenovirus early region 4 is essential for normal stability of late nuclear RNAs. J Virol 63: 624–630
Sarnow P, Hearing P, Anderson CW, Halbert DN, Shenk T, Levine AJ (1984) Adenovirus early region 1B 58,000-dalton tumor antigen is physically associated with an early region 4 25,000-dalton protein in productively infected cells. J Virol 49: 692–700
Schlehofer JR, Ehrbar M, zur Hausen H (1986) Vaccinia virus, herpes simplex virus, and carcinogens induce DNA amplification in a human cell line and support replication of a helpervirus dependent parvovirus. Virology 152: 110–117
Senepathy P, Tratschin J, Carter BJ (1984) Replication of adeno-associated virus DNA. Complementation of naturally occurring *rep*⁻ mutants by a wild-type genome or an *ori*⁻ mutant and correction of terminal deletions. J Mol Biol 179: 1–20
Siegel G, Bates RC, Berns KI, Carter BJ, Kelly DC, Kurstak E, Tattersall P (1985) Characteristics and taxonomy of *Parvoviridae*. Intervirology 23: 61–73
Snyder RO, Samulski RM, Muzyczka N (1990a) In vitro resolution of covalently joined AAV chromosome ends. Cell 60: 105–133
Snyder RO, Im D-S, Muzyczka N (1990b) Evidence for covalent attachment of the adeno-associated virus Rep protein to the ends of the AAV genome. J Virol 64: 6204–6213
Srivastava A, Lusby EW, Berns KI (1983) Nucleotide sequence and organization of the adeno-associated virus 2 genome. J Virol 45: 555–564
Srivastava CH, Samulski RJ, Lu L, Larsen SH, Srivastava A (1989) Construction of a recombinant human parvovirus B19: adeno-associated virus 2 (AAV) DNA inverted terminal repeats are functional in an AAV-B19 hybrid virus. Proc Natl Acad Sci USA 86: 8078–8082
Straus SE, Sebring ED, Rose JA (1976a) Concatemers of alternating plus and minus strands are intermediates in adenovirus-associated virus DNA synthesis. Proc Natl Acad Sci USA 73: 742–746
Straus SE, Ginsburg HS, Rose JA (1976b) DNA-minus temperature-sensitive mutants of adenovirus type 5 help adenovirus-associated virus replication. J Virol 17: 140–148
Thimmappaya B, Weinberger C, Schneider RJ, Shenk T (1982) Adenovirus VAI RNA is required for efficient translation of viral mRNAs at late times after infection. Cell 31: 543–551
Tratschin J-D, Miller IL, Carter BJ (1984a) Genetic analysis of adeno-associated virus: properties of deletion mutants constructed in vitro and evidence for an adeno-associated virus replication function. J Virol 51: 611–619
Tratschin J-D, West MHP, Sandbank T, Carter BJ (1984b) A human parvovirus, adeno-associated virus, as a eukaryotic vector: transient expression and encapsidation of the prokaryotic gene for chloramphenicol acetyltransferase. Mol Cell Biol 4: 2072–2081
Tratschin JD, Miller IL, Smith MG, Carter BJ (1985) Adeno-associated virus vector for high-frequency integration, expression, and rescue of genes in mammalian cells. Mol Cell Biol 5: 3251–3260
Tratschin J-D, Tal J, Carter BJ (1986) Negative and positive regulation in *trans* of gene expression from adeno-associated virus vectors in mammalian cells by a viral Rep gene product. Mol Cell Biol 6: 2884–2894
Trempe JP, Carter BJ (1988a) Regulation of adeno-associated virus gene expression in 293 cells: control of mRNA abundance and translation. J Virol 62: 68–74
Trempe JP, Carter BJ (1988b) Alternate mRNA splicing is required for synthesis of adeno-associated virus VP1 capsid protein. J Virol 62: 3356–3363
Trempe JP, Mendelson E, Carter BJ (1987) Characterization of adeno-associated virus rep proteins in human cells by antibodies raised against rep expressed in *Escherechia coli*. Virology 161: 18–28
Tsao J, Chapman MS, Agbandjo M, Keller W, Smith K, Wu H, Luo M, Smith TJ, Rossman MG, Compans RW, Parrish CR (1991) The three-dimensional structure of canine parvovirus and its functional implications. Science 25: 1456–1464
Vincent KA, Moore GK, Haigwood NL (1990) Replication and packaging of HIV envelope genes in a novel adeno-associated virus vector system. *Vaccine* 90: 353–359
Walsh CE, Liu JM, Young NS, Nienhuis AW, Samulski RJ (1991) Gene transfer and high level of expression of a human globin gene mediated by a nove adeno-associated virus (AAV) vector. American Clinical Research/American Federation of Clinical Investigation, abstract
West MHP, Trempe JP, Tratschin J-D, Carter BJ (1987) Gene expression in adeno-associated

virus vectors: the effects of chimeric mRNA structure, helper virus, and adenovirus VAI RNA. Virology 160: 38–47

Winocour E, Callahan MF, Huberman E (1988) Perturbation of the cell cycle by adeno-associated virus. Virology 167: 393–399

Wondisford FE, Usala SJ, DeCherney GS, Castern M, Radovick S, Gyves PW, Trempe JP, Kerfoot BP, Nikodem VM, Carter BJ, Weintraub BD (1988) Cloning of the human thyrotropin β-subunit gene and transient expression of biologically active human thyrotropin after gene transfection. Mol Endocrinol 2: 32–39

Yacobson B, Koch T, Winocour E (1987) Replication of adeno-associated virus in synchronized cells without the addition of a helper virus. J Virol 61: 972–981

Yacobson B, Hrynko TA, Peak MJ, Winocour E (1989) Replication of adeno-associated virus in cells irradiated with UV light at 254 nm. J Virol 63: 1023–1030

Yalkinoglu AO, Heilbronn R, Burkle A, Schlehofer JR, zur Hausen H (1988) DNA amplification of adeno-associated virus as a response to cellular genotoxic stress. Cancer Res 48: 3123–3125

The Baculovirus-Infected Insect Cell as a Eukaryotic Gene Expression System

M. J. FRASER

1 Introduction

Baculovirus-infected insect cell cultures are established as an easily manipulated eukaryotic system for highly efficient expression of gene products. The system takes advantage of several unique attributes of this virus group, including highly active late gene promoters, the capacity for insertion of large fragments of foreign DNA, replication competence of the resulting recombinants, and the relative ease of handling both the insect cell cultures and the viruses.

Department of Biological Sciences, University of Notre Dame, Notre Dame, Indiana 46556, USA

Current Topics in Microbiology and Immunology, Vol. 158

This review will attempt to give an overview of the use of baculoviruses as expression vectors, focusing on the *Autographa californica* nuclear polyhedrosis virus as the prototype virus system. The available evidence for proper and improper post-translational processing, cellular localization, and antigenicity of various protein products will be discussed. Several of the currently employed expression vectors are described, and several of the key techniques used in generating baculovirus recombinants will be reviewed. For additional information on baculovirus expression vector constructs, handling the viruses and cell cultures, and analyzing recombinants the reader is referred to previous reviews by MILLER et al. (1986), SUMMERS and SMITH (1987), LUKOW and SUMMERS (1988), KANG (1988), MILLER (1988), CAMERON et al. (1989), VLAK and KEUS (1990) and LUCKOW (1991).

2 The Biology of Baculoviruses

The insect-pathogenic nuclear polyhedrosis viruses (NPVs) are members of subgroup A of the family Baculoviridae. These DNA viruses replicate within the nuclei of susceptible insect cells and have a complex, essentially biphasic replication cycle that generates two infectious forms, extracellular budded virus (ECV), and occlusion bodies (OBs). The two infectious forms of the virus are genotypically identical but phenotypically distinct, each serving a vital function in the survival of the virus in host insect populations. The OBs are an environmentally stabilized form of the virus that function to initiate the primary infection within the gut of host insects, while the ECVs serve to disseminate the virus between cells in the insect host and are employed during all manipulations of the virus in vitro.

The NPVs may be further subdivided based upon the arrangement of nucleocapsids in the occluded virions within OBs. The M-type viruses occlude enveloped single and multiple nucleocapsids, while the S-type viruses occlude only enveloped single nucleocapsids. The prototype virus for the M-subtype is AcMNPV (MATTHEWS 1982), while the S-subtype is represented by the *Heliothis zea* NPV (HzSNPV; CORSARO and FRASER 1987a).

The following discussions are meant to serve as a general introduction for those unfamiliar with baculovirus replication. For more detailed discussion of baculovirus biology and genetics the reader may consult recent reviews (DOERFLER and BOHM 1986; GRANADOS and FEDERICI 1986; BLISSARD and ROHRMANN 1990).

2.1 Viral Life Cycle

During the first phase of the lytic infection (0–20 h) the rod-shaped nucleocapsids are assembled within the nucleus of infected cells in pockets of the virogenic matrix (FRASER 1986a). Electron-microscopic observations suggest that the bulk of the rod-shaped capsid sheath assembles prior to incorporation of the circular 128-kb virus genome, and that extension of the capsid may occur during genome packaging, permitting larger than unit length genomes to be incorporated (FRASER 1986a). Infectious ECVs are released by budding from the virus-modified cell surface. The cell membrane-derived envelope of these budded virions has peplomeric extensions of a viral encoded glycoprotein, gp64, that is a major neutralizing antigen and is apparently involved in adsorption of the virions to host cell surfaces (VOLKMAN et al. 1984).

Since the functions of nucleocapsid assembly and budding of infectious ECV are necessary for survival of the virus both in vivo as well as in vitro, they are sometimes referred to as essential functions (FRASER 1986b). In contrast, functions carried out in the second phase are necessary for survival of the virus in vivo, but not in vitro, and are considered to be nonessential or conditionally essential.

The second phase becomes apparent at about 20 h post infection (p.i.) and continues until the cells expire. Nucleocapsids remaining within the nucleus at the beginning of the second phase are sequestered in de-novo-synthesized envelopes and then encapsulated within OBs (HUGHES 1972; STOLTZ et al. 1973; FRASER 1986a). The OBs are formed by the assembly of a paracrystalline matrix composed of a single protein called polyhedrin (SUMMERS and SMITH 1976). The encapsulation of virions in OBs protects the virus during desiccation, and helps stabilize the virus against extremes of heat and cold. Upon ingestion by a suitable host insect, the OB matrix is dissolved in the basic pH of the insect gut juices, and the embedded virions are released to initiate infection of the midgut epithelial cells (see FAULKNER 1981 for a review).

In a broad sense, the second phase of NPV replication accomplishes functions that are common among many insect viruses that infect the larval stages of insects that undergo complete metamorphosis. This phase can be likened to encapsulation in prokaryotes or sporulation in lower eukaryotes, and serves to insure survival of the viruses for prolonged periods within soil or on leaf surfaces. Similar strategies for environmental survival are employed by the closely related granulosis viruses (GV; Baculoviridae Subgroup B; MATTHEWS 1982), the insect-pathogenic poxviruses (entomopoxviruses; ARIF 1984) and the insect-pathogenic cytoplasmic polyhedrosis viruses (CPV; PAYNE and MERTENS 1983).

2.2 Expression of Baculovirus Genes

As with most viruses, the temporal regulation of baculovirus gene expression is a tightly controlled cascade initiating with the immediate early genes (alpha class), followed by the delayed early genes (beta class), both of which precede DNA replication (KELLY and LESCOTT 1981; CARSTENS et al. 1979; MILLER et al. 1983b). Alpha genes are first expressed following penetration and uncoating of the virus in the cell nucleus (GUARINO and SUMMERS 1986). These genes do not require any previously coded viral proteins for expression (KELLY and LESCOTT 1981), since purified viral DNA is capable of initiating and completing the entire replication cycle (see below). The beta class of genes are dependent upon the alpha class products for expression (KELLY and LESCOTT 1981; MILLER et al. 1983b; GUARINO and SUMMERS 1986, 1987).

Following the initiation of genome replication (5–7 h p.i; CARSTENS et al. 1979), the late genes (gamma class) are expressed (CARSTENS et al. 1979; KELLY and LESCOTT 1981). These genes presumably encode functions related to virus structure and assembly. The expression of late genes corresponds with the release of ECV from the infected cells (CARSTENS et al. 1979; WOOD 1980; KELLY and LESCOTT 1981).

Most of the alpha, beta, and gamma gene products produced during the first phase of the replication cycle are essential to baculovirus viability. Such essential processes as viral specific gene activation, degradation of the host cell genome, replication of the viral genome, assembly of nucleocapsids, and transport and budding of virions from the cell surface are all mediated by these first-phase genes. Up to this point the baculovirus replication process superficially resembles that of many DNA animal viruses.

Baculovirus replication differs from other DNA animal virus groups in having a fourth temporal class of expressed genes. At about 20 h p.i. the release of ECV dramatically declines, expression of many alpha, beta, and gamma proteins is relatively reduced but not necessarily entirely eliminated (CARSTENS et al. 1979; WOOD 1980; GUARINO and SUMMERS 1986), and the delta class proteins predominate (CARSTENS et al. 1979; WOOD 1980; KELLY and LESCOTT 1981). The delta genes comprise the second phase of the baculovirus replication cycle. Among the products generated in AcMNPV infected cells during this second phase are those proteins involved in the construction of nuclear localized de novo envelopes, the 29-kDa polyhedrin protein that forms the paracrystalline matrix of the OBs, the 10 K protein whose function is unclear (SMITH et al. 1983a; VLAK et al. 1988), and the 34 K protein which is a component of the OB envelope (WHITT and MANNING 1988). The temporally regulated cascade of gene activation insures that maximal expression of foreign genes under control of the very late polyhedrin gene promoter, or other delta gene promoters, will not begin until well after virus replication is completed.

As mentioned before, these very late genes are considered nonessential or conditionally essential, and are superfluous to effective replication of the virus in cell cultures. However, mutations that abolish OB formation severely limit the replication potential of the virus under natural conditions and the virus cannot be maintained in the natural host unless infectious OBs are formed. In this respect the use of the polyhedrin substitution vector effectively introduces a form of biological containment for the baculovirus expression vectors.

2.3 Polyhedrin and Other Late Genes

Because of its relative abundance during NPV infections and its importance in the natural survival of the virus, polyhedrin was the first baculovirus protein to be identified and studied in detail. The conditionally essential nature of the polyhedrin gene made it an ideal first candidate for genetic engineering in the NPV system.

Polyhedrins of different baculoviruses are biochemically similar. The proteins isolated from both MNPVs and SNPVs are similar in size, isoelectric points, solubility properties, and immunoreactivity (see VLAK and ROHRMANN 1985 for a comprehensive review). Polyhedrins are readily solubilized at a pH of 9.5–10.5 or greater. All polyhedrins sequenced to date have a calculated molecular weight between 28 and 29 kDa, pI between 5.3 and 6.5, and are immunologically cross-reactive.

The polyhedrin proteins of NPVs are abundantly expressed at late times during infection. Polyhedrin is estimated to be as much as 25% of total infected cell protein (PENNOCK et al. 1984) and up to 50% of the stainable protein expressed during the second phase of replication (LUCKOW and SUMMERS 1988; MILLER 1988). The protein accumulates to estimated levels of 1 mg per $1–2 \times 10^6$ cells (LUCKOW and SUMMERS 1988), principally in the form of 50 to 100 OB particles per infected cell. These OBs are the predominant structure of wild-type infected cells (2–5 μm in diameter) and can be observed easily with a light microscope. Moreover, polyhedrins are one of the very few viral proteins that are produced throughout the entire second phase of NPV infections, permitting accumulation of the protein over an extended period (usually 48 h or more).

Many early studies of baculovirus replication suggested the nonessential nature of polyhedrin. Electron-microscopic observations of baculovirus maturation had always demonstrated that OBs and related second-phase structures form well after nucleocapsid assembly and ECV release (KNUDSON and HARRAP 1976; ADAMS et al. 1977). Polyhedrin synthesis in AcMNPV-infected cell cultures always became obvious from 20 h p.i. and reached a maximum between 36 and 48 h p.i. In contrast, infectious ECV release was detected by 8 h p.i. and became maximal around 18 h p.i. Ultimately, the nonessential nature of polyhedrin was proven by experimental elimination of the polyhedrin gene through in vitro mutagenesis and gene transfer (SMITH et al. 1983b).

The polyhedrin genes from a number of viruses have been sequenced in entirety (HOOFT VAN IDDEKINGE et al. 1983; IATROU et al. 1985; LEISY et al. 1986a). The size and structure of polyhedrin genes is quite similar, and there is a 5′ untranslated A + T rich leader sequence that is strikingly similar among all polyhedrin genes and begins with a 12–bp consensus sequence surrounding the mRNA start site (ROHRMANN 1986; HOWARD et al. 1986).

Another major late protein in AcMNPV-infected cells is the p10 protein. The exact function of this protein is still in doubt, but it appears to be involved in a nonessential way in the assembly or structuring of the occlusion body envelope (QUANT-RUSSELL et al. 1987; VAN DER WILK et al. 1987). Recent studies have demonstrated that p10 is present as a component of the fibrous material (FM), a nuclear inclusion present at late times in infection and involved in formation of the OB envelope (VLAK et al. 1988). Deletion of the p10 coding region does not abolish FM formation (WILLIAMS et al. 1989), nor does it inhibit OB envelope formation (VLAK et al. 1988). Fusion of p10 with *lacZ* abolishes OB envelope formation, rendering the OBs more labile to solubilization (VLAK et al. 1988). In addition, the p10/*lacZ* fusion protein can be localized to the FM inclusion, but the exact function of p10 in development of FM or OB envelopes is unknown (VLAK et al. 1988).

3 Engineering Recombinant Baculoviruses

The potential utility of the polyhedrin gene to effect high-level expression of foreign genes in baculovirus-infected insect cells was recognized independently by several investigators (SMITH et al. 1983b; MILLER et al. 1983a; MAEDA et al. 1985).

In the first report of successful application of the baculovirus as an expression vector (SMITH et al. 1983c), the recombinant viruses incapable of producing OBs were distinguished from wild-type virus by a simple plaque assay in a manner analogous to the previously established detection methods for FP mutants (FRASER and HINK 1982). Baculovirus plaques are microscopic, and best viewed with a dissecting microscope at between 20 × and 70 × magnification. Illumination of the infected cell monolayers with obliquely directed light causes an opalescent appearance in wild-type virus plaques due to the presence of numerous refractive OBs in each infected cell. Those plaques lacking OBs (OB negative), or with significant reductions in the number of OBs per infected cell (i.e., FP mutants) are discernably less opalescent.

An alternative strategy was to construct *lacZ*-gene fusions at the natural *Bam*HI (+ 171 bp) within the polyhedrin coding region (PENNOCK et al. 1984). Co-transfections of recombinant plasmid and wild-type viral DNAs resulted in recombinant viruses that were selected based upon their ability to form blue plaques in the presence of the colorimetric indicator, X-gal, and their inability to form OBs.

The capacity for production of foreign genes in infected insect larvae was demonstrated by MAEDA et al. (1985). In this study, the polyhedrin coding region of *Bombyx mori* MNPV (BmMNPV) was localized using cDNAs prepared from mRNAs isolated from the fat bodies of infected silkworm larvae. The availability of the BM-N established cell line allowed propagation and manipulation of the virus to generate recombinants. In this case, however, the expression of the polyhedrin-controlled gene product was demonstrably more efficient in the larval system than in cell cultures.

Another major late protein gene of AcMNPV, p10, has been manipulated in a manner similar to the polyhedrin gene (VLAK et al. 1988). The gene was localized to the *Eco*RI-P fragment by hybridization of abundantly expressed late mRNAs to viral gene fragments (SMITH et al. 1983a). Subsequent sequencing of the *Eco*RI-P fragment positioned the gene and it's promoter (KUZIO et al. 1984). A sequence similar to the consensus 5′ mRNA start site of polyhedrins is located upstream of the p10 gene as well (KUZIO et al. 1984; ROHRMANN 1986), suggesting this is a common recognition sequence for hyperexpression of late genes. Since this gene does not provide a selective phenotype, *lac*Z fusions were performed at the unique *Bgl*II site at + 153 bp, and recombinants were selected for blue-plaque morphology (VLAK et al. 1988). The *lac*Z recombinants were replication competent, as expected for a nonessential late gene, and were capable of forming OBs as well.

Engineering the p10 region for expression of genes necessarily requires either selection of clear plaques following cotransfection of a p10/*lac*Z recombinant with the p10 expression vector construct, or construction of transfer plasmids having both *lac*Z and the gene of choice under control of a hyperexpressed promoter. Such double-promoter expression vectors have been constructed and have proved effective (see below).

Promoters for other late genes that are either essential or nonessential may be utilized for expression vector construction. One example of the use of an essential gene promoter is the study of HILL-PERKINS and POSSEE (1990). In this instance they utilized the promoter for the major core protein gene (WILSON et al. 1987), a delta class gene product that seems to be associated with the packaging of viral DNA (TWEETEN et al. 1980; BUD and KELLEY 1980). The strategy involved duplication of the promoter for the core protein gene next to the *lac*Z gene inside the polyhedrin gene region. Substitution of the constructed gene for the nonessential polyhedrin gene generated virus expressing *lac*Z maximally between 8 h and 18 h p.i., as would be expected for a delta class promoter (HILL-PERKINS and POSSEE 1990). These authors point out that the use of promoters from earlier temporal classes may have some advantages for production of proteins requiring extensive post-translational modifications, or to express insect specific toxins and hormones at earlier times in the infected insect than with the p10 or polyhedrin promoters.

3.1 The Polyhedrin Promoter

Examination of the 5′ regions from a number of polyhedrins has led to some interesting observations relative to baculovirus late gene promoter structure and sequences necessary for hyperexpression of polyhedrins and p10 proteins. The first reports of the polyhedrin gene sequences had indicated the presence of apparent TATA- and CAT-like promoter signals (HOOFT VAN IDDEKINGE et al. 1983; MAEDA et al. 1985; IATROU et al. 1985). However, the comparative analysis of upstream flanking sequences for a number of polyhedrin genes by ROHRMANN (1986) demonstrated little conservation of sequence and position of the TATA- and CAT-like regions. Instead, there was remarkable similarity in the length and AT content of the 5′-nontranslated leader region. In addition, the comparison identified a consensus sequence surrounding the mRNA start site at nucleotide −49 in a number of polyhedrin genes (LEISY et al. 1986b). This consensus sequence, between nucleotide −43 and −54 (5′AATAAGTATTTT3′) is apparently essential for high level expression of the polyhedrin gene.

Several sequences within the nontranslated leader are clearly involved in optimal expression of the gene. MATSUURA et al. (1987) demonstrated that deletions in the leader region past −14 from the ATG start codon adversely affect the expression of genes fused to the polyhedrin promoter, while deletion of the entire coding region and 3′ flanking sequences does not appreciably alter their expression. POSSEE and HOWARD (1987) demonstrated that deletions in the 5′ flanking region from −69 to −92 did not affect levels of *lac*Z production by polyhedrin/*lac*Z recombinant viruses. However, deletion of the region from −47 to −56 (encompassing the transcriptional start site and the Rohrmann consensus sequence) resulted in significantly reduced levels of *lac*Z-specific transcript. They described the functional limits of the polyhedrin promoter as between −49 and −69. Similarly, RANKIN et al. (1988) demonstrated that removal of the region from −1 through to the *Eco*RV site at −92 bp reduces the expression of the chloramphenicol acetyl transferase (*CAT*) gene 1000-fold in a transient expression assay. Interestingly, positioning of this leader-containing region fused to the *CAT* gene in reverse orientation with respect to the flanking viral sequences increased the levels of *CAT* activity twofold. Successive substitution of a 10-bp *Hin*dIII linker for sequences within the 5′ flanking region from −83 to −1 was used to assess the relative importance of leader sequences on promoter function. The most dramatic reductions in *CAT* expression (nearly 1000-fold) were obtained with substitutions in the region from −42 to −60, once again demonstrating the importance of the consensus 5′ mRNA initiation signal in polyhedrin gene expression. OOI et al. (1989) suggested the importance of the nontranslated leader in optimal transcriptional initiation.

A sequence similar to the consensus 5′ start site of polyhedrins is located upstream of the p10 gene as well (KUZIO et al. 1984; ROHRMANN 1986), suggesting this is a common recognition sequence for hyperexpression of very late genes.

3.2 Available Polyhedrin-Based Baculovirus Transplacement Vectors

The baculovirus expression vector systems now available include those derived from AcMNPV, BmMNPV, and HzSNPV. No expression vectors are yet available for the other major subgroup of baculoviruses, the granulosis viruses, owing to a lack of suitable in vitro systems for their propagation and manipulation. The AcMNPV-infected *Spodoptera frugiperda* cell line system (SMITH et al. 1983c; PENNOCK et al. 1984; SUMMERS and SMITH 1987) has received the most attention due to the relative ease of handling, and the ready availability of the cell line, media, and virus. Several vector constructs are currently available for the manipulation of the AcMNPV virus polyhedrin gene. The reader is referred to other recent reviews for a more complete description of available vectors and their activities (LUCKOW and SUMMERS 1988; LUCKOW 1991).

To date, the pAc373 construct has been the vector used most often. This vector is superior to pAc380 for expression of human interleukin-2 (IL-2; SMITH et al. 1985), probably because it retains the consensus transcription initiation signal and much of the 5′ leader sequence of the polyhedrin gene. The pAc373 vector contains a deletion from the *Bam*HI site (+171) of the wild-type AcMNPV polyhedrin sequence, through the ATG start codon, and 5′ untranslated leader sequence, to −8 from the polyhedrin messenger RNA cap site. Similar vectors have also been constructed by MATSUURA et al. (1986). The insertion of a *Bam*HI linker (pAcRP6) or a multiple cloning site (pAc373) allows insertion of genes in the untranslated leader region. In both cases, the natural *Bam*HI site at +171 is fused to the *Bam*HI site of the linker region to complete the constructions.

Similar manipulations resulted in the construction of the pAc610 vector, also used by several investigators. This vector differs from pAc373 in having a more extensive inserted polylinker sequence after nucleotide −7, and a more extensive deletion of the polyhedrin coding domain to nucleotide +670 near the terminus of the polyhedrin gene (LCUKOW and SUMMERS 1988). The pAcRP18 and pAcYM1 (MATSUURA et al. 1987) vectors contain a *Bam*HI linker fused at positions −1 and +1, respectively, pEV55 (MILLER et al. 1986) is constructed with a polylinker at position +1. pEVmod eliminates redundant sites in pEV55 at the pUC8/AcMNPV sequence junctions (WANG et al. 1991), and pEVmXIV substitutes a modified polyhedrin promoter, P_{XIV} (RANKIN et al. 1988; OOI et al. 1989) for the wild-type polyhedrin promoter (WANG et al. 1991). All of these vectors leave all, or nearly all, of the nontranslated leader sequence intact. Both fused and nonfused expression vectors are available for BmMNPV as well (S. Maeda, personal communication).

The pAcCL29-1 and pAcCL29-8 vectors are derivatives of pAcYM1 that contain an M13 origin of replication which permits production of single-stranded DNA in the presence of a helper phage (LIVINGSTON and JONES 1989). These vectors facilitate site-directed mutagenesis for analysis of expressed genes. A similar strategy was used by HASEMANN and CAPRA (1990) by incorporating the F1 origin of replication into pAc360-derived transfer vectors containing murine immunoglobulin heavy and light chain regions.

From the first descriptions of expression vectors it was noted that in many cases the expression of genes fused in-frame with the polyhedrin coding sequences is much greater than those placed near the start of transcription, or even near the start of translation. Levels of both human beta-interferon and bacterial beta-galactosidase in AcMNPV were quite high if expressed as in-frame fusions with the polyhedrin coding sequences rather than as nonfused proteins (SMITH et al. 1983c; PENNOCK et al. 1984). LUCKOW and SUMMERS (1988) indicated that in-frame fusions of 30 amino acids or more of polyhedrin to several protein genes increased the levels of expressed protein and RNA. Apparently sequences in the 5′ amino-terminal region of the polyhedrin coding domain are also involved in optimal expression of the polyhedrin gene.

More recently constructed vectors, such as the pVL941 vector (LANFORD 1988) or its derivatives pVL1392 and pVL1393 (mentioned in LUCKOW 1991), contain an alteration of the polyhedrin ATG start codon to ATT, allowing genes to be inserted with their own ATG start codons at a *Bam*HI site downstream of the normal polyhedrin translation initiation sites. The pVL941 vector provided twofold higher levels of SV40 large T antigen expression than the pAc373 vector (LANFORD 1988).

A similar vector, p36C, was constructed by mutagenesis of the ATG start codon to ATC in the polyhedrin fusion vector pAc360 (PAGE 1989). Genes are inserted at the retained *Bam*HI site at +33, and levels of expressed protein product are reportedly 5 times greater than with pAc373.

There is ample evidence that the 3′ terminal regions of the polyhedrin gene have little, if any, effect on expression of the gene. Deletions of this region failed to introduce any appreciable alterations in levels of expressed RNA or protein (MATSUURA et al. 1987; POSSEE and HOWARD 1987).

The relative levels of expression obtained with each of these vectors depends to a great extent on the gene being expressed. While optimal expression of similar genes is effected when all of the 5′ leader and some polyhedrin coding domain are intact (e.g. LANFORD 1988), the actual amount of a given protein product generated is most dependent on the gene being expressed (LUCKOW and SUMMERS 1988).

Recently, a vector based upon a synthetic promoter, P_{syn} (WANG et al. 1991), has been constructed from comparisons of polyhedrin and p10 (see below) promoter regions. Analyses of *CAT* gene expression have demonstrated that this promoter is less efficient than the wild-type polyhedrin promoter, but can be used at numerous alternative regions in the virus genome without duplicating existing baculovirus sequences, thus avoiding potential instability (WANG et al. 1991).

3.3 Available p10-Based Baculovirus Transplacement Vectors

Expression vectors utilizing the promoter for p10 have been constructed and demonstrated to be effective for foreign gene expression. The overall selection strategy relies upon the prior construction of a p10-*lac*Z recombinant virus to serve as recipient. p10 expression vector recombinants generate white plaques from these p10-*lac*Z recipient viruses.

The vector pAcUW1 provides a unique *Bgl*II site at position +1 in the p10 gene (WEYER et al. 1990) and essentially mimics the promoter structure of pAcYM1. The effectiveness of this vector was demonstrated using both the *lac*Z gene and the polyhedrin gene. The p10-*lac*Z fusion gene was also used as a selectable marker for polyhedrin-based expression vectors. The pAcUW2 constructs (WEYER et al. 1990) allow insertion of genes under control of the p10 promoter and just upstream of a functional polyhedrin gene. Recombinant virus generated from this sort of vector would be capable of forming OBs while expressing the gene of choice. Such constructs could have advantages for scale-up in insect larvae or for genetically improved biological insecticides.

The pAcAS2 vector (VLAK et al. 1990) was constructed with a pUC19 multiple cloning site at position +1 in the p10 gene. Subsequent addition of a *Drosophila* hsp70 promoter-driven *lacZ* gene to produce the vector pAcAS3 provides a selectable marker for detecting recombinant expression viruses. The utility of the pAcAS3 vector was demonstrated by expression of cauliflower mosaic virus gene 1 (VLAK et al. 1990).

3.4 Practical Considerations

Among the advantages of the baculovirus system are the relative ease of handling of both the virus and cell cultures and the speed with which one can engineer and isolate recombinants for analysis. It is not unreasonable to have plaque purified recombinant viruses for analysis within 6 weeks from successful cloning of the gene to be expressed (either an intronless genomic or cDNA copy of the gene is advised; see below).

The basic technology of gene replacement or allelic replacement is employed for genetic engineering of the baculovirus genome. Essentially, this involves the transfection of viral and recombinant plasmid DNAs into susceptible insect cell cultures and the identification of recombinants based on the absence of the large intranuclear OB inclusions. Most of the procedures commonly employed in working with baculoviruses have been assembled as a manual by SUMMERS and SMITH (1987).

The transfection of lepidopteran insect cells with viral or plasmid DNAs is most conveniently accomplished using the $CaPO_4$ coprecipitation procedure originally developed for mammalian cells (GRAHAM and VAN DER EB 1973; BURAND et al. 1980; POTTER and MILLER 1980a; CARSTENS et al. 1980). Modifications of this precipitation procedure have been employed on occasion, but are generally not as reliable. The modification typically employed for transfection of *Drosophila* cell lines (DINOCERA and DAWID 1983) is much less effective for lepidopteran cells. Alternative transfection procedures such as the polybrene method (KAWAI and NISHIZAWA 1984) are less effective than the $CaPO_4$ method for lepidopteran cells and baculorius DNAs. Lipofection (BRL) has also been used (SHIU et al. 1991), and in our hands this procedure is effective but has not represented a significant improvement over $CaPO_4$ technique (M. J. Fraser, unpublished). Electroporation is very effective for a *S. frugiperda* cells in particular (MANN and KING 1989) and other lepidopteran cell lines in general (M. J. Fraser, unpublished), but may be inconvenient for general use.

In general, the relative efficiency of the $CaPO_4$ transfection procedure is related to a number of factors influencing the precipitation reaction (GRAHAM et al. 1980). In addition to proper formation of the precipitate, the most important factors include the quality of the viral and plasmid DNAs and the cell line employed. For the generation of baculovirus recombinants, the greatest efficiencies and most consistent results are obtained with CsCl gradient purified viral and plasmid DNAs. Dialysis of the DNAs against 0.1 × SSC is recommended over ethanol precipitation as a final step in the isolation (CORSARO and FRASER 1989).

The IPLB-SF21AE cell line (VAUGHN et al. 1977) and derivatives (i.e., Sf-9; SUMMERS and SMITH 1987) is an ideal recipient cell line for genetic manipulations of AcMNPV by transfection. In contrast, the TN-368 (HINK 1970) cell line is not as receptive to transfection with the same preparations of AcMNPV. The UND-K derivative of the IPLB-HZ 1075 cell line (CORSARO and FRASER 1987b) is suitable for manipulation of HzSNPV while the parent cell line, IPLB-HZ 1075 and several other cloned derivatives are less receptive (CORSARO et al. 1989). The BM-N cell line (MAEDA et al. 1985) is suitable for genetic manipulations with BmMNPV.

Once the viral and plasmid DNAs are transfected into the recipient cell line, recombination between viral sequences on the replacement vector and the same regions of the viral genome takes place (SMITH et al. 1983b, c; PENNOCK et al. 1984; MAEDA et al. 1985). This recombination event is presumably mediated by cellular factors, but is also influenced by the extent of unmodified flanking viral sequences in the transplacement vector (for a review see LUCKOW and SUMMERS 1988) Most polyhedrin-based vectors currently employed retain a minimum of 7 kb flanking viral DNA. With such vectors, the recombination efficiency is of the order of 0.1%–5%.

The quality of the viral DNA is also critical in deriving recombinant viruses. We have noted for some time that supercoiled viral DNA is 5 times more infectious in transfections of *S. frugiperda* cell cultures (CORSARO and FRASER 1989a), but gives a reduced yield of recombinant virus compared to the nicked circular or linear forms (unpublished observations). KITTS et al. (1990) took advantage of this fact in engineering a virus, AcRP6-SC, that contains a unique *Bsu*36I site within the polyhedrin gene. Linearization of the viral DNA prior to cotransfection with the polyhedrin transfer vector containing the gene to be expressed decreased the overall infectivity of the transfected viral DNA, but increased the relative yield of recombinant virus to between 6% and 32% of the progeny virus. Similar results were obtained for linearization of viral DNA at the p10 locus using p10 gene transfer vectors. These results suggest that only circularized viral DNA molecules may replicate, and that addition of the appropriate transfer vector provides a means for selective repair within a given locus by recombination with the digested viral DNA. This effectively increases the relative proportion of progeny virus that are recombinant.

GOSWAMI and GLAZER (1991) report success using plasmid DNA purified by passage through Quiagen columns (Quiagen, Studio City, CA) to transfect Sf-9 cells that had been previously infected with wild-type AcMNPV. The advantage of this approach is the elimination of lengthy procedures for preparation of viral and plasmid DNAs. However, direct comparisons with alternative preparations and estimates of relative efficiency of recombinant virus recovery were not made.

3.5 Detecting OB-Negative Recombinant Plaques

Selection of recombinant virus can be accomplished in a number of ways. The most commonly employed technique for distinguishing recombinants takes advantage of the reduced refractivity of plaques that produce fewer OBs than the wild-type virus. This detection method had been previously perfected through analysis of the spontaneous FP mutants of baculoviruses (FRASER and HINK 1982). The ability to visually detect the recombinant plaques depends upon optimization of the plaque assay methodology. The fact that some researchers experience difficulty detecting recombinants in this way indicates these factors are not trivial.

The initial seeding cell density, overlay formulation, and incubation conditions can all influence the detection of OB-negative plaques (FRASER and HINK 1982). Because the virus spreads relatively slowly in the monolayer, the cells must be seeded at a density allowing growth of the monolayer for several days. Early death of the monolayer will preclude plaque formation or hamper detection of recombinants. A low initial cell density will not yield a dense enough monolayer for sufficient localized cell death to permit detection of recombinant plaques. The optimal cell seeding density differs for a given cell line, and is largely dependent upon the cell doubling time. If detection of OB-negative plaques is difficult, plaque assays of an OB-negative mutant control virus should be performed with varying cell seeding densities to optimize conditions for detection. We have found optimal plaque formation with *S. frugiperda* cells occurs with seeding densities of between 1.1 and 1.25×10^3 cells per mm^2 of available plate surface.

Staining viable cells for 15–20 min with a small volume of sterile 0.01%–0.05% neutral red in PBS added over the agarose overlay and viewing the infected monolayer 4–8 h later enhances the detection of OB negative plaques significantly. The alternative strategy of adding neutral red directly to the overlay mixture (KNUDSON 1979) has not been as successful in our hands.

Checking the putative recombinant plaques at 200 X magnification with an inverted microscope is highly recommended prior to picking. Potential false positives, such as the spontaneous FP mutants that also generate plaques of reduced refractivity, can occasionally be found in preparations of wild-type virus. These FP mutants can be distinguished from recombinants by the presence of at least some OBs in infected cells of the plaque (see FRASER 1986b, for review). Since these mutants can be amplified upon continued propagation of the virus in cell cultures, baculovirus DNAs used for transfections to generate recombinants should be prepared from plaque-purified stocks of wild-type virus that has undergone fewer than five passages in cell culture, or from OBs purified from peroral-infected insect larvae.

Several cycles of plaque purification are recommended to insure purity of the recombinant virus. Multiple potential recombinants should be isolated for analysis at this stage. As a final check, the putative recombinants should be analyzed with restriction enzymes, and possibly Southern hybridization, to insure the desired DNA fragment is present. As with any virus, the baculovirus genome is a dynamic entity (KUMAR and MILLER 1987) subject to rearrangements (BURAND and SUMMERS 1982), insertional mutagenesis by host cell sequences (POTTER and MILLER 1980b; FRASER et al. 1983; MILLER and MILLER 1982) and other less apparent types of mutations. Checking recombinants by restriction enzyme analysis and other means prior to amplification insures at least that major alterations of the viral genome have not taken place.

3.6 Alternative Selection Schemes

Alternative detection methods have occasionally been employed with varying success. The ability to adsorb red blood cells can help identify recombinant virus-infected cells that are producing viral hemagglutinins such as the influenza or parainfluenza envelope glycoproteins (KURODA et al. 1986, 1987; VAN WYKE COELINGH et al. 1987).

Limiting dilution and DNA dot-blot hybridization has been employed with success to detect and purify recombinant virus (FUNG et al. 1988; PEN et al. 1989; GOSWAMI and GLAZER 1990). This method allows selective amplification of recombinant virus, avoids potential selection of false-positive plaques, and works well with even low recombination frequencies. Hybridization plaque lift assays may also be employed (VILLAREAL and BERG 1977; MILLER et al. 1986; SUMMERS and SMITH 1987; JEANG et al. 1987b), but these assays meet with varying success, and additional manipulations will be necessary to optimize conditions.

Newer vectors are available that take advantage of the selective *lacZ* marker gene in plaque assays. These vectors rely on alternative promoters to express the *lacZ* gene, while the polyhedrin promoter is reserved for expression of the gene of choice. Blue viruses are easily detected after addition of the colorimetric substrate X-gal to the agarose overlay (PENNOCK et al. 1984). One example of this approach is the pJV(NheI) vector (VIALARD et al. 1990), a derivative of pAc373 that contains a p10 promoter expressing beta-galactosidase in opposite orientation next to a polyhedrin promoter region reconstituted to include sequences from the polyhedrin gene up to +33, with the normal ATG start codon altered to ATT. VIALARD et al. (1990) report that at least 70% of *lacZ*-positive viruses also express the gene of choice.

Another such vector, pAcDZ1, constructed by ZUIDEMA et al. (1990) utilizes a chimeric gene formed from the *Drosophila melanogaster* 70 K heat shock gene promoter (hsp70), the beta-galactosidase protein coding domain, and the SV40 early region termination signals. In this case, the polyhedrin gene and the chimeric *lacZ* gene are opposed, with the SV40 termination signals separating them. The levels of expressed protein product obtained with pAcDZ1 were similar to those obtained with the conventional polyhedrin-based expression vectors (ZUIDEMA et al. 1990). A similar strategy was employed for expression vectors based on the p10 promoter (VLAK et al. 1990).

LacZ-selectable vectors are also commercially available from Invitrogen Corporation (San Diego, CA) as part of the MaxBac[R] Baculovirus Expression System kit. The only potential problem with the *lacZ* co-expression approach may be that the increased size of these transfer vectors can make cloning certain genes for expression difficult.

4 Expression of Foreign Genes in Baculovirus-Infected Insect Cells

The attractiveness of the baculovirus-infected insect cell expression system rests on the fact that it allows for extremely efficient expression of protein products (averaging 1–10 μg per 10^6 cells) in a higher eukaryote cell system. Maximal expression of the protein occurs after the essential phase of virus replication, potentially allowing expression of gene products that may be cytotoxic (MILLER 1988). The expression system itself provides a level of biological containment (MAEDA 1987; MILLER 1988) because introduction of foreign genes into the polyhedrin promoter abolishes a function necessary for survival of the virus under natural conditions.

The baculovirus is capable of accommodating a large excess of sequences without appreciable effect on replication efficiency. The largest inserted fragment to date is approximately 10 kb (CARBONELL et al. 1985). Larger insertions up to 15 kb have been observed, based upon analyses of spontaneous mutations (FRASER 1986a). However, the stability of larger inserts still needs to be explored. It is possible that tandemly duplicated sequences or sequences flanked by inverted repeats will be unstable in this recombination competent system.

Another attractive feature of the system is the possibility of altering the viral genome at several locations due to the presence of multiple nonessential genes and intergenic mutable regions. As the location of mutable regions and additional late genes becomes known, more areas will be available for manipulation. This should permit incorporation of several highly expressed genes at multiple locations throughout the genome. In this regard, the manipulations of EMERY and BISHOP (1987) demonstrated the feasibility of incorporating more than one polyhedrin promoter in a single virus. Both proteins, the native polyhedrin protein and the recombinant lymphocytic choriomeningitis virus N protein, accumulated to substantial levels in the infected cells. However, the relative amount of polyhedrin produced by the recombinant virus was somewhat reduced compared to the control AcMNPV. Whether these reduced levels were significant was not determined.

Coinfection of insect cells with a number of different recombinant viruses is an alternative strategy that has proven effective for simultaneous expression of multiple gene products. ST ANGELO et al. (1987) demonstrated the effective coexpression of three influenza polymerase subunit genes (PA, PB1, and PB2) in Sf-9 cells coinfected with three recombinant viruses each expressing an individual gene. Expression of these genes occurs within the same cell, since protein complexes are formed between two of the subunits (PB1 and PB2). Immunoglobulin heterodimers could be formed by coinfecting Sf-9 cells with two recombinant viruses carrying either heavy- or light-chain murine immunoglobulin genes (HASEMANN and CAPRA 1990).

The expression of multiple genes by coinfection offers opportunities for studying the interaction of overexpressed proteins in a eukaryotic cell environment devoid of significant background cellular protein synthesis. Expression of multiple mammalian genes in insect cells provides an environment potentially free of interfering proteins. Coexpression of the mouse p53 protein and the SV40 large T antigen in recombinant-infected insect cells resulted in typical complex formation between the two recombinant proteins, similar to that which would occur in SV40-infected mouse cells (O'REILLY and MILLER 1988). Co-infection with recombinants expressing $pp90^{rsk}$ and $pp60^{v-src}$ resulted in activation of the serine-specific protein kinase activity of *rsk* through tyrosine phosphorylation by *src* (VIK et al. 1990).

The overexpression of eukaryotic gene products in insect cell lines provides several advantages over conventional prokaryotic expression systems. Post-translational modifications such as signal peptide cleavage, N-linked and O-linked glycosylation, additional proteolytic cleavages (although perhaps not identical to those of mammalian cells), and proper cellular compartmentalization of protein products (i.e., membrane localization, extracellular secretion, cytosolic localization, nuclear localization) all occur in baculovirus-infected insect cells. Both intrachain and interchain disulfide bridge formation have been observed (GEISE et al. 1989). These is now ample evidence for RNA splicing as well (CHISHOLM and HENNER 1988; JEANG et al. 1987a; IATROU et al. 1989), although the preferential splicing of sites other than those preferred in mammalian cells has been observed (JEANG et al. 1987a).

Some advantages have been noted for baculovirus-expressed proteins compared to expression of the same proteins in bacteria. Insoluble protein aggregates that frequently form upon overexpression of genes in prokaryotes may not occur during expression with baculoviruses. JEANG et al. (1987b) point out that the p40x protein of HTLV-I forms aggregates when expressed in *Escherichia coli*. In contrast, insoluble aggregates are not formed by the same protein expressed in the baculovirus expression vector system, even though each infected cell produced 50–100 times more protein than the bacterial system. Baculovirus-produced Rap1A protein was both soluble and biochemically active as opposed to the same protein produced in *E. coli* (QUILLIAM et al. 1990). HSEIH et al. (1989) report 5 times greater levels of expression for rat liver Yb_1 glutathione S-transferase than in *E. coli*.

4.1 Proteolytic Cleavages in Baculovirus-Infected Insect Cells

There is ample evidence that signal peptides are correctly cleaved from a number of diverse membrane-bound or secreted proteins expressed in baculovirus infected insect cells. Amino-terminal analysis has confirmed correct signal peptide cleavage for the human alpha-interferon (MAEDA et al. 1985), human gastrin-releasing peptide (LEBACQ-VERHEYDEN et al. 1988), human IL-2 (SMITH et al. 1985), mouse IL-3 (MIYAJIMA et al. 1987), human glucocerebrosidase (MARTIN 1988), human T-cell immune activation gene *Act*-2 (LIPES et al. 1988), *Phaseolus vulgaris* beta-phaseolin (BUSTOS et al. 1988), Sindbis virus E1 envelope protein (OKER-BLOM and SUMMERS 1989), and immunoglobulin heavy- and light-chain proteins (HASEMANN and CAPRA 1990). Correct cleavage of signal peptides is inferred from activity assays, transport and secretion, and size analysis on SDS-polyacrylamide gels for a number of other proteins. Examples include human beta-interferon (SMITH et al. 1983c), human erythropoietin (WOJCHOWSKI et al. 1987), the influenza hemagglutinins (POSSEE 1986; KURODA et al. 1986), parainfluenza type 3 hemagglutinin-neuraminidase (VAN WYKE COELINGH et al. 1987), the HIV envelope glycoprotein gp160 (HU et al. 1987; RUSCHE et al. 1987; COCHRAN et al. 1987), *hst-1* transforming protein (MIYAGAWA et al. 1988), leech antistasin (HAN et al. 1989), G1 and G2 glycoproteins of Rift Valley fever virus (SCHMALJOHN et al. 1989), the S glycoprotein of bovine coronavirus (YOO et al. 1990), human granulocyte-macrophage colony stimulating factor (CHIOU and WU 1990), and the alpha subunit of human chorionic gonadotropin (NAKHAI et al. 1991). The insect cells did not recognize and cleave a signal peptide for a bacterial protein, the *Bacillus anthracis* protective antigen (IACONO-CONNORS et al. 1990).

Some types of post-translational proteolytic and protein modifying processes are apparently lacking or different in insect cells. Detailed comparisons of the processing of human gastrin releasing peptide precursor in baculovirus-infected insect cells and a mammalian lung cancer cell line revealed significant differences in several proteolytic cleavages (LEBACQ-VERHEYDEN et al. 1988). No carboxypeptidase B-like activity, trypsin-like endopeptidase activity, or peptidyl glycine alpha-amidating monooxygenase activity was detected following processing in Sf-9 cells. Instead, peptides that were similar in size to several of those present in the mammalian cell line were produced by proteases with previously undefined specificities.

WATHEN et al. (1989a) report that there are significant differences in the cleavage of the F_0 precursor glycoprotein of human respiratory syncytial virus (RSV) produced in insect cells. The extent of cleavage F_0 to F_1 and F_2 was markedly reduced compared to the vaccinia-expressed protein in Vero cells, and a secondary cleavage of the F_1 molecule was seen in the baculovirus-expressed protein. However, the proteins produced in insect cells were capable of inducing a neutralizing antibody response in cotton rats, although at lower levels than with protein from RSV-infected cells. Similarly, the F_0 glycoprotein of measles virus was only partially cleaved when expressed in insect cells, suggesting these cells are deficient in this endoproteolytic activity, while monkey kidney cells cleave this protein efficiently (VIALARD et al. 1990). However, VIALARD et al. (1990) indicate that an insect cell line from *Trichoplusia ni* was able to cleave the F_0 more efficiently, confirming that there may be considerable variation in proteolytic activities among established lepidopteran cell lines. The baculovirus-expressed F_0 glycoprotein of humam parainfluenza type 3 was not cleaved at all into the F_1 and F_2 subunits in insect cells (RAY et al. 1989), but the recombinant protein was effective in inducing a protective immune response in hamsters.

Although the study of LEBACQ-VERHEYDEN et al. (1988) detected no trypsin-like endoprotease activity operating at the lysine residues in the sequence Gly-Lys-Lys-Ser, a trypsin-like endoprotease activity and a carboxypeptidase N activity were inferred from correct maturation of the influenza (fowl plague) virus hemagglutinin (KURODA et al. 1986, 1987). In this sutdy the majority of recombinant hemagglutinin produced in *S. frugiperda* cells was uncleaved precursor HA, but the cleaved products HA1 and HA2 were also apparent. The correct cleavage at the carboxy-terminal Arg and Gly residues the sequence Lys-Lys-Arg-Lys-Lys-Arg-Gly is essential for activation of the hemolytic fusion activity of the influenza hemagglutinin. Since such an activity could be isolated from homogenates derived from recombinant infected *S. frugiperda* cells, correct cleavage was inferred (KURODA et al. 1986, 1987).

This observation was believed significant because many other vertebrate glycoproteins have similar cleavage sites, and might be expected to be efficiently processed as well. However, no endoproteolytic cleavage of precursor HA to HA1 and HA2 was observed with the hemagglutinin of a human influenza virus (POSSEE 1986). Similarly, cleavage of the gp160 envelope glycoprotein of the HTLV-IIIB isolate of HIV to gp120 and gp41 was not detected by immunoblotting of total infected cell proteins separated on SDS-polyacrylamide gels (RUSCHE et al. 1987), or by immunoprecipitation of baculovirus-produced LAV gp160 (HU et al. 1987). Apparently certain trypsin-like recognition sites will be effectively cleaved while others will not. The exact reason for this discrepancy is not clear.

Finally, a 5′ terminal fragment encoding the C, M, and E structural proteins and the NS1 and NS2a nonstructural proteins of dengue virus 4 was expressed using the pAc373 vector (ZHANG et al. 1988). In this case, cotranslational processing of the polyprotein sequences resulted in apparently authentic NS1 and E glycoproteins, and such processing is believed to require the specific acitivity of cellular proteolytic enzymes. However, the levels of E and NS1 produced were only 25% of that generated in dengue virus-infected primate cells. Even so, the baculovirus-infected cell lysates were capable of inducing immunity to dengue virus in mice, apparently as a result of antibodies generated to the recombinant NS1 glycoprotein (ZHANG et al. 1988).

4.2 N-Linked Glycosylation and Fucosylation

Most glycoproteins that have been expressed in baculovirus-infected cells are immunologically active and, in some cases, have been effective in inducing protective immunity. However, the published reports on expressed glycoproteins reflect considerable variability concerning the exact nature and extent of glycosylation that can occur on various proteins. Simple size comparisons of glycoproteins produced in both baculovirus-infected insect cells and mammalian cells often seem to reveal differences in the extent of glyosylation (KURODA et al. 1986, 1987, 1990; POSSEE 1986; HU et al. 1987; RUSCHE et al. 1987; COCHRAN et al. 1987; STEINER et al. 1988; GREENFIELD et al. 1988; FURLONG et al. 1988; DOMINGO and TROWBRIDGE 1988; KRISHNA et al. 1989; BAILEY et al. 1989; GRABOWSKI et al. 1989; GEORGE et al. 1989; QUELLE et al. 1989; JOHNSON et al. 1989; OKER-BLOM et al. 1989; GERMANN et al. 1990; YOO et al. 1990; SANCHEZ-MARTINEZ and PELLETT 1991; VAN DRUNEN LITTLE-VAN DEN HURK et al. 1991). In most cases these discrepancies have been demonstrated, through further analyses, to result from differences in type and extent of glycosylation of the protein between insect cells and mammalian cells.

There are at least two reports in which baculovirus-produced glycoproteins are not glycosylated at all. Neither the human multidrug transporter P-glycoprotein (GERMANN et al. 1990) nor the recombinant extracellular domain of the human nerve growth factor (VISSAVAJJHALA and ROSS 1990) were glycosylated.

Tunicamycin treatment (SMITH et al. 1983c, 1985; BUSTOS et al. 1988; HASEMANN and CAPRA 1990; SANCHEZ-MARTINEZ and PELLETT 1991; JANSEN et al. 1991; NIIKURA et al. 1991a) and labeling with radiolabeled mannose (SMITH et al. 1983c, 1985; HU et al. 1987; COCHRAN et al. 1987; BAILEY et al. 1989) have both effectively demonstrated N-linked glycosylation. Enzymatic digestions have also been employed to characterize the glycosylation processes. N-linked glycosylation has been demonstrated in baculovirus-infected insect cells through the use of N-glycanase (RUSCHE et al. 1987; WOJCHOWSKI et al. 1987; MARTIN et al. 1988; GRABOWSKI et al. 1989; QUELLE et al. 1989; CHIOU and WU 1990), endoglycosidase-F (COCHRAN et al. 1987; DOMINGO and TROWBRIDGE 1988; WATHEN et al. 1989a, b; GERMANN et al. 1990; DESPRES et al. 1991) and Glycopeptidase-F (STEINER et al. 1988; JOHNSON et al. 1989; VAILARD et al. 1990). The presence of high mannose oligosaccharides is also indicated through the use of endoglycosidase-H (TARENTINO and MALEY 1974; MARTIN et al. 1988; WEBB et al. 1989; GRABOWSKI et al. 1989; DOMINGO and TROWBRIDGE 1988; GREENFIELD et al. 1988; JOHNSON et al. 1989; OKER-BLOM and SUMMERS 1989; FELLEISEN et al. 1990; DESPRES et al. 1991), and by adsorption of expressed glycoproteins to concanavalin A-Sepharose columns (STEINER et al. 1988; BUSTOS et al. 1988; QUELLE et al. 1989), and the trimming of high mannose residues in indicated by conversion of some glycoproteins to endoglycosidase-H resistance upon secretion (COCHRAN et al. 1987; JARVIS and SUMMERS 1989; JOHNSON 1989; SISSOM and ELLIS 1989; DESPRES et al. 1991). However, the extent of this processing is apparently reduced from that in mammalian cells (JARVIS and SUMMERS 1989).

JARVIS and SUMMERS (1989) also employed inhibitors of the mannose oligosaccharide processing pathway to identify the enzyme activities present in baculovirus infected Sf-9 insect cells. Both castanospermine and N-methyl-deoxynojirimycin apparently inhibited processing of high mannose residues, suggesting the presence of glucosidases I and/or II in Sf-9 cells. Analyses with inhibitors of mannosidase I and II activity were less convincing, but the presence of some mannosidase activity is strongly implied by conversion of glycoproteins to endoglycosidase-H resistance.

The complex class of N-linked oligosaccharides have not been generally observed for proteins produced in insect cells. Levels of galactosyl and sialyl transferases had been reported to be negligible in insect cells (BUTTERS et al. 1981) and terminal sialic acid residues are not generally observed in insect glycoproteins (KURODA et al. 1986, 1987, 1990).

There are, however, an increasing number of reports indicating that some proteins may undergo complex glycosylations. Several studies have produced baculovirus-expressed glycoproteins similar in size to their authentic counterparts (JARVIS and SUMMERS 1989; WHITEFLEET-SMITH et al. 1989; WEBB et al. 1989; RAY et al. 1989; KLAIBER et al. 1990; SCHMALJOHN et al. 1989, 1990; RODEWALD et al. 1990; KOENER and LEONG 1990; NAGY et al. 1990; SHIU et al. 1991; NIIKURA et al. 1991a; JANSEN et al. 1991). Since many of these authentic proteins are known to have complex oligosaccharides, the similarity in size on gel electrophoresis suggests similar patterns of glycosylation in the baculovirus-expressed proteins.

Fucosylation of N-linked oligosaccharides has also been recently confirmed. WATHEN et al. (1991) analyzed N-linked oligosaccharides with bovine epididymis alpha-fucosidase and detected the presence of a fucosylated trimannosyl structure. This confirms the primary observation by KURODA et al. (1990) implying the presence of fucosyltransferases in insect cells.

DAVIDSON et al. (1990) provided the first evidence of complex glycosylation of a protein expressed in insect cells. Anion-exchange liquid chromatography mapping of glycopeptidase F-released oligosaccharides demonstrated that approximately 40% of baculovirus-expressed human plasminogen released bisialo-biantennary complex-type carbohydrate. Subsequent analysis (DAVIDSON and CASTELLINO 1991) showed that the appearance of complex carbohydrates on recombinant human plasminogen was dependent on the time post infection. Early in the infection (0–20 h p.i.) approximately 96% of all oligosaccharides are of the high mannose type, while 92% of the those released from recombinant plasminogen between 60 and 90 h p.i. were of the comlex type. These results suggest that the normal glycosylation patterns of the insect cells may be altered in response to the presence of recombinant human plasminogen. Similar analyses of baculovirus-expressed plasminogen in the *Mamestra brassicae* insect cell line IZD-MB0503 revealed that while normal insect cell proteins may not contain N-linked complex oligosaccharides the glucosyltransferase activities required for assembly of complex oligosaccharides can be effected upon infection with the human plasminogen-expressing baculovirus recombinant (DAVIDSON and

Castellino 1991). A possible explanation for the apparent inducibility of complex glycosylation in insect cells during baculovirus infection is that (alpha 1,2)-D-mannosidase-like enzyme activity is relatively low in IPLB-SF21AE cells but is somehow stimulated upon infection with either wild-type AcMNPV or a plasminogen-expressing recombinant (Davidson et al. 1991).

The apparent variability between reports of oligosaccharides processing in insect cells should not be unexpected. Patterns of N-linked glycosylation can vary with the cell line employed, culture media and supplements used, length of time post infection, and most importantly, the protein itself (Parekh et al. 1989). Many of these variables differ between individual reports of glycoprotein expression in baculovirus-infected lepidopteran cells.

Not all cell types will perform glycosylation in the same way (Parekh et al. 1989). The lepidopteran cell lines used for baculovirus-mediated expression vary among different laboratories. For example, many reports use the Sf-9 subclone of the IPLB-SF21AE cell line, while others use the parental SF21AE cells. At least one paper reports the use of a Sf158 subcloned cell line derived from an unspecified *S. frugiperda* cell line. Kuroda et al. (1986, 1990) and Shiu et al. (1991) report a *S. frugiperda* cell line, but do not specify the source.

Given the fact that changes in the physiological condition can alter patterns of glycosylation (Parekh et al. 1989) the media and culture conditions used for analysis of glycosylated patterns may influence the extent and timing of glycosylation in a given cell line. Media formulations employed for baculovirus-mediated expression range from TNM-FH with or without serum supplement, Grace's with or without serum, TC-100 with or without serum, EX-CELL 400 serum-free medium (JRH Biosciences, Lexana, KS), or other serum-free formulations (see below).

For any given combination of cell type, medium, culture conditions, time of harvesting, and protein expressed, the results are quite consistent. However, glycosylation patterns seen for a specified protein under a given set of conditions may not be generally seen for all glycoproteins under the same conditions. The patterns and timing of glycosylation are highly dependent upon the protein being expressed (Parekh et al. 1989).

While most glycosylation events in insect cells may not be identical to those of mammalian cells, they are evidently sufficient to allow extracellular transport of secretory proteins and cells surface presentation of membrane glycoproteins. They also do not preclude formation of tertiary and quaternary structures essential for biological activity or antigenic potential (e.g., fowl plague virus HA protein, Kuroda et al. 1986; parainfluenza virus type 3 HN protein, van Wyke Coelingh et al. 1987; hepatitis B virus S antigen 22 nm particle formation, Kang et al. 1987; Scully and Kang 1988; vesicular stomatitis virus G protein, Bailey et al. 1989; Rift Valley fever virus G1 and G2 proteins, Schmaljohn et al. 1989; Newcastle disease virus HN protein, Niikura et al. 1991a).

The accumulated evidence suggests that baculovirus recombinant glycoproteins can be expected to be immunologically and functionally similar to the native proteins. In some cases they may even be superior to mammalian cell devied protein in inducing an immune response (e.g., HSV-1 glycoprotein D, KRISHNA et al. 1989), while in other cases they may be less effective (e.g., bovine herpesvirus 1 glycoprotein gIV, VAN DRUNEN LITTEL-VAN DEN HURK et al. 1991). The effectiveness of insect cell-produced proteins as therapeutic agents remains a topic of debate, and can only be resolved through further experimentation.

4.3 O-Linked Glycosylation

There is less evidence in the published literature for O-linked glycosylation in this expression system. However, analysis of baculovirus-expressed pseudorabies glycoprotein gp50 in infected Sf-9 cells demonstrated the presence of O-linked polysaccharides by digestions with endo-alpha-*N*-acetylgalactosaminidase (THOMSEN et al. 1990). Subsequent analysis of the released disaccharide with bovine testis beta-galactosidase and jack bean beta-galactosidase suggested a beta 1–3 linkage between the monosaccharides. Enzyme assays confirmed the presence of N-acetylgalactosaminyl transferase and beta-1, 3-galactosyl transferase activities in Sf9 cells (THOMSEN et al. 1990). Similar analyses were performed by WATHEN et al. (1991) in examining the oligosaccharide structures on a respiratory syncytial virus chimeric FG protein.

4.4 Fatty Acid Acylation

Both palmitoylation and myristoylation have been detected in recombinant-infected insect cells. Palmitoylation has been verified through labeling of a baculovirus-expressed SV40 large T antigen (LANFORD 1988), Ha-*ras* p21 (PAGE et al. 1989), and human transferrin receptor (DOMINGO and TROWBRIDGE 1988) with ^{3}[H]-palmitic acid. Similarly, myristoylation of the baculovirus-expressed preS-S polypeptide of the hepatitis B surface antigen (HBsAg; PERSING et al. 1987; LANFORD et al. 1988), the p17 *gag*-related protein of HIV (OVERTON et al. 1989), and the FIV *gag* precursor protein (MORIKAWA et al. 1991) was verified through labeling with ^{3}[H]-myristic acid. These reports confirm that the necessary enzymes for fatty acid acylation are present in baculovirus infected insect cells.

4.5 Nuclear Transport

A number of nuclear localized proteins have been expressed in the baculovirus system. One example is the human c-*myc* protein (MIYAMOTO et al. 1985). Baculovirus-expressed c-*myc* was phosphorylated, efficiently transported to the nucleus, and was found tightly associated with the nuclear matrix. These properties are identical to those of the native protein. Other phosphorylated nuclear proteins that have been expressed are the *Kruppel* gene product of *Drosophila* (OLLO and MANIATIS 1987), the presumptive transposase of the Maize Ac transposon (HAUSER et al. 1988), and the mouse c-*ets*-1 protooncogene (CHEN 1990). Additional nuclear proteins successfully expressed include the human Ku autoantigen (ALLAWAY et al. 1990), human *fos* protein (TRATNER et al. 1990), and the *Egr*-1 transcription factor (RAGONA et al. 1991).

The expression of the qa-1F activator protein of *Neurospora crassa* demonstrated the effectiveness of employing a baculovirus late gene promoter for producing potentially cytotoxic gene products. The qa-1 activator protein was apparently toxic when expressed in either *E. coli* or *Saccharomyces cerevisiae*, but could be expressed when introduced into a baculovirus vector (BAUM et al. 1987). The recombinant protein had the same DNA-binding specificity as the native form, and was useful in determining the location of binding sites within the qa gene cluster.

4.6 Expression of Viral Nonstructural Gene Products

A number of viral nuclear proteins have also been successfully expressed. Recombinant polyomavirus large T antigen was antigenically similar to the mammalian form and displayed polyomavirus origin-specific DNA binding (RICE et al. 1987). The SV40 large T antigen was effectively phosphorylated and displayed origin-specific DNA binding as well (MURPHY et al. 1988). In addition, the SV40 large T antigen displayed an ATPase activity comparable to that of the mammalian cell-derived counterpart, indicating retention of significant elements of the native protein conformational structure (MURPHY et al. 1988). A similar analysis of SV40 T-antigen expression by LANFORD (1988) detected palmitylation, glycosylation, and oligomerization of T-antigen produced in baculovirus-infected insect cells. All of these modifications of recombinant large T antigen are found in the native protein isolated from mammalian cells (KLOCKMANN and DEPPERT 1983).

Transactivating gene products from a number of mammalian viruses have been expressed and characterized using the baculovirus system. These include the herpes simplex virus (HSV) alpha-TIF regulatory protein (KRISTIE et al. 1989) and trans-activator Vmw65 (CAPONE and WERSTUCK 1990), the human T-cell leukemia virus type I (HTLV-I) $p40^{x}$ trans-activator (NYUNOYA et al. 1988; JEANG et al. 1987b), the bovine leukemia virus (BLV) $p34^{tax}$, and the hepatitis B virus X protein (SPANDAU et al. 1991).

The transactivating $p40^{x}$ protein of HTLV-I is phosphorylated (NYUNOYA et al. 1988; JEANG et al. 1987b) and can transactivate an HTLV-I LTR indicator target plasmid in recombinant-infected insect cells (JEANG et al. 1987b). In contrast, the recombinant transactivating protein, *tat*, of the human immunodeficiency virus (HIV) type 1 was not phosphorylated and its biological activity could not be assessed directly in infected Sf-9 cells (JEANG et al. 1988a). Instead, Sf-9 cells infected with the *tat* recombinant virus were fused with a mammalian cell line containing an integrated HIV LTR-CAT reporter plasmid. The detection of CAT activity following fusion of the baculovirus-infected insect cells and the transformed mammalian cells confirmed the biological activity of the recombinant *tat* protein (JEANG et al. 1988a). Activity of recombinant HTLV-I and HIV-1 *tat* proteins was evident in the absence of de novo cellular protein synthesis (JEANG et al. 1988b). These studies established cell fusion as an effective strategy for assessing the biological acitivity of recombinant proteins synthesized in insect cells in a mammalian cell environment.

The baculovirus-infected insect cell can provide large quantities of easily purified mammalian virus proteins for research purposes. Recombinant dengue virus 4 structural and nonstructural proteins may be useful in defining epitopes that are protective from those that are immunopathogenic (ROTHMAN et al. 1989). Recombinant baculoviruses have been utilized extensively to facilitate genetic and biochemical analyses of the function and activities of animal virus gene products. These include the adenovirus type 2 DNA polymerase (WATSON and HAY 1990), polyoma virus middle T antigen (FORSTOVA et al. 1989), poliovirus $3B^{VPg}$, $3C^{pro}$, and $3D^{pol}$ (NEUFELD et al. 1991), the human papillomavirus type 18 E6 protein (GROSSMAN et al. 1989), the herpes simplex virus UL5, UL8, UL9, and UL52 gene products (OLIVO et al. 1988, 1989) and DNA polymerase (MOARCY et al. 1990), and several different HIV-1 *pol* gene products (HU and KANG 1991). While the baculovirus-produced Epstein-Barr virus alkaline deoxyribonuclease will be useful for biochemical analyses, it may also be an effective reagent for diagnosis of nasopharyngeal carcinoma (BAYLIS et al. 1991).

4.7 Expression of Virus Structural Proteins

Numerous mammalian virus capsid protein genes have been expressed for potential use as diagnostic or vaccine reagents. A cDNA encoding the major rotavirus capsid protein VP6 was efficiently expressed and formed trimeric capsid subunits and oligomeric assemblies (ESTES et al. 1987). All of the native immunoreactive determinants were conserved as assessed by reactivity with a battery of monoclonal antibodies. In addition, antiserum prepared against baculovirus-produced VP6 was highly specific for the viral protein, and lacked reactivity with mammalian proteins (ESTES et al. 1987). These results suggested potential advantages for use of baculovirus-produced proteins in diagnostic test kits.

The expression of other rotavirus gene products has facilitated characterization of some of their properties or functions. Examples include the outer capsid protein VP4 of the rhesus rotavirus which was characterized as the virus-neutralizing antigen and hemagglutinin (MACKOW et al. 1989), the bovine rotavirus VP1 (COHEN et al. 1989), and the Simian rotavirus nonstructural phosphoprotein product of gene 11 (WELCH et al. 1989). Passive immunization of suckling mice could be effected by inoculating female mice with baculovirus-expressed VP4 (MACKOW et al. 1990).

Several bluetongue virus (BTV) gene products have been analyzed following expression (INUMARU and ROY 1987; INUMARU et al. 1987; URAKAWA and ROY 1988; FRENCH et al. 1989; OLDFIELD et al. 1990; THOMAS et al. 1990). The recombinant VP2 protein of BTV serotype 10 was capable of inducing neutralizing antibody titers in both mice and rabbits, and might be useful as a subunit vaccine (INUMARU and ROY 1987), while recombinant VP5 was incapable of eliciting a neutralizing antibody response (MARSHALL and ROY 1990). The recombinant BTV VP3 and VP7 group-specific antigens are effective in detecting several BTV serotypes by indirect ELISA and may be useful as diagnostic reagents (INUMARU et al. 1987; OLDFIELD et al. 1990). Baculovirus expression of NS1, another group-specific antigen, demonstrated that this protein forms the virus-specific tubules seen during BTV-10 infections (URAKAWA and ROY 1988; MARSHALL et al. 1990), while the NS2 gene product is the structural component of the virus inclusion bodies characteristic of BTV-10 infections (THOMAS et al. 1990). Both of these structures are formed in the recombinant baculovirus-infected insect cells as well.

Several other recombinant baculovirus-produced viral proteins show promise of being effective diagnostic antigens. Examples include Hantaan virus structural proteins (SCHMALJOHN et al. 1988), hepatitis virus core and surface antigens (TAKEHARA et al. 1988), the hepatitis A VP1 capsid protein (HARMON et al. 1988), a dengue 4 virus core/PreM fusion protein (MAKINO et al. 1989), the Lassa fever virus nucleocapsid protein (BARBER et al. 1990), the hepatitis C virus core protein (CHIBA et al. 1991), human papillomavirus (HPV) 6b E2 gene product (SEKINE et al. 1989), the L2 open reading frames of both HVP 6b and 11 (ROSE et al. 1990), the human parvovirus B19 VP1 and VP2 structural proteins (BROWN et al. 1990), rabies virus nucleoprotein (PREHAUD et al. 1990), and the gp57-65 of Marek's disease virus (NIIKURA et al. 1991b), which may also be useful as a vaccine reagent.

With respect to producing antigens for diagnostic assays, ROTA et al. (1990) point out that one advantage of producing influenza A and B nucleoprotein antigens in recombinant baculovirus-infected insect cells is that the proteins need not be purified prior to use in ELISA assays. They were unable to detect antibodies in human, mouse, ferret, rabbit, or chimpanzee sera that reacted with control baculovirus-infected insect cells. MILLS and JONES (1990) produced the p24 core protein of HIV and were able to enrich the recombinant protein to greater than 90% purity from soluble fractions of infected insect cells in a simple two-step procedure of ammonium sulfate precipitation and gel filtration. The resulting preparation was capable of detecting p24 antibodies in all AIDS patients' sera tested and exhibited no background reactivity with serum from a noninfected individual. Similar low levels of background reactivity were reported by DEVASH et al. (1990) in Western blots to detect antibodies towards baculovirus-produced HIV *rev* and *vif* gene products in ARC and AIDS patients' sera.

The HIV *gag* region and a *gag*/*pol* segment including the entire *gag* region and 65% of the *pol* gene were each expressed in this system (MADISEN et al.

1987). The results indicate that initial proteolytic processing of *gag* from the 55-kDa precursor polypeptide to the 40-kDa precursor is apparently mediated by cellular proteases common to both insect and mammalian cells. Subsequent proteolysis of the p40 polypeptide to mature viral proteins p24, p18, and p14 is mediated by a *gag*-specific protease activity encoded by the *pol* region. The recombinant *gag* polypeptides were specifically recognized by AIDS patient serum suggesting they could serve as useful diagnostic reagents (MADISEN et al. 1987).

The $Pr55^{gag}$ precursor protein of HIV-1 assembles into virus-like particles (VLPs) in recombinant baculovirus-infected insect cells (GHEYSEN et al. 1989). The particles are targeted to the cell membrane and mutagenic analyses established that myristoylation of the amino-terminal glycine residue is essential for budding of spherical particles from the cell. Similar analyses were performed on simian immunodeficiency virus $Pr57^{gag}$ (DELCHAMBRE et al. 1989).

VLPs that were myristoylated and secreted were also obtained upon expression of the feline immunodeficiency virus (FIV) *gag* precursor protein. Inclusion of the FIV protease gene with the *gag* precursor gene allowed processing of the *gag* precursor and no VLPs were released (MORIKAWA et al. 1991). VLPs formed during expression of the bovine immunodeficiency virus precursor, $Pr53^{gag}$, could be processed in vitro utilizing the protease activity from NP-40-lysed preparations of purified BIV virions (RASMUSSEN et al. 1990). These studies point to the utility of these baculovirus-produced precursor *gag* VLPs for structural studies and for analysis of inhibitors of viral-specific protease activity, a necessary function in the maturation of infectious virions.

The ability to produce noninfectious VLPs using baculovirus expression vectors offers the potential for generating safe vaccine preparations. Noninfectious empty capsids are produced during coexpression of human parvovirus B19 structural proteins VP1 and VP2, or upon expression of VP2 alone (BROWN et al. 1991; KAJIGAYA et al. 1991). However, only particles containing VP1 were capable of inducing a neutralizing antibody response immunized rabbits (KAJIGAYA et al. 1991). Correct processing of capsid precursor polyprotein P1 into VP0, VP1, and VP3 occurred during expression of the complete coding region of poliovirus type 3. The capsid proteins assembled into noninfectious VLPs that were able to induce neutralizing antibodies in immunized mice (URAKAWA et al. 1989). Expression of the foot-and-mouth disease virus P1-2A region in the presence of the 3C protease yielded correct cleavage of the P1-2A polyprotein into capsid proteins 1AB, 1C and 1D but the efficiency of aggregation into particles was low (ROOSIEN et al. 1990).

Cowpea mosaic virus genes were also successfully expressed in baculovirus-infected cells, and at higher levels than that obtained for the same genes in bacteria. Correct proteolytic processing of precursor polypeptides by the expressed recombinant 24 K viral protease also occurred (VAN BOKHOVEN et al. 1990).

4.8 Other Proteins of Interest

The uniqueness of the baculovirus-infected insect cell environment, coupled with the characteristics of high-level expression, have induced many investigators to use the system for biochemical investigations of protein structure and function. Tyrosine hydroxylase, implicated in the pathogenesis of neuropsychiatric disorders, has been expressed for analyses of its biochemical and pharmacological properties (GINNS et al. 1988; FITZPATRICK et al. 1990). Human 5-lipoxygenase (FUNK et al. 1989) and human terminal transferase (CHANG et al. 1988) were expressed at relatively high levels and retained their enzymatic activities in insect cells. The function and structure of various domains of the insulin receptor have also been examined (ELLIS et al. 1988; HERRERA et al. 1988; VILLALBA et al. 1989; PAUL et al. 1990).

A variety of protein kinase genes or kinase domains of membrane receptors have been shown to retain their respective activity and specificity (WEDEGAERTNER and GILL 1989; PATEL and STABEL 1989; BRICKEY et al. 1990). Several G protein-coupled receptors were localized in the insect cell membrane and retained their activity and selectivity (PARKER et al. 1991; REILANDER et al. 1991).

A few insect-related proteins have been expressed, including the firefly luciferase gene (HASNAIN and NAKHAI 1990) and attacin (GUNNE et al. 1990), and two plant proteins, patatin and beta-phaseolin, have also been efficiently expressed (ANDREWS et al. 1988; BUSTOS et al. 1988).

The general application of gene transfer and expression in baculoviruses has renewed interest in their potential as microbial insecticides. Several different approaches involving expression of insect-specific toxins have been examined by various investigators. CARBONELL et al. (1988) describe an unsuccessful attempt to use the insect-specific scorpion toxin. A similar approach using the insect-specific scorpion neurotoxin gene of a different species proved more successful (STEWART et al. 1991) and reduced the time required for the virus to kill the insect host.

MAEDA (1989b) describes the use of a diuretic hormone from *Manduca sexta* to engineer recombinant BmMNPV which kill infected larvae 20% faster than the wild-type virus in injection experiments. Introduction of a recombinant baculovirus expressing the juvenile hormone esterase (JHE) gene from *Heliothis virescens* proved somewhat effective in reducing feeding behavior of first-instar *T. ni* larvae, but due to the instability of the JHE in vivo, was less effective on later stages (HAMMOCK et al. 1990).

MARTENS et al. (1990) engineered AcNPV to express the *Bacillus thuringiensis* crystal endotoxin gene *cryIA*(b) in place of the polyhedrin gene. The CryIA(b) protein produced in baculovirus-infected *S. frugiperda* cells formed crystals that were toxic to a susceptible insect species.

A newly isolated gene encoding a mite neurotoxin shows promise of being useful for enhancing efficacy of recombinant baculoviruses for control of insect pests. Recombinant baculoviruses producing the toxin were effective in paralyzing fifth-instar larvae within 2 days of injection, while larvae infected with the same dose of wild-type virus continued feeding (TOMALSKI and MILLER 1991). As other insect-specific toxins are identified, the prospects for utilizing recombinant baculoviruses as pest control agents will continue to improve.

5 Scale-Up Considerations

The lepidopteran cell lines commonly employed for growth and expression of baculovirus vectors are relatively hardy. Cultures can be maintained at room temperature if desired, although the optimal growth temperature is 27° to 29° C. Insect cell culture media are not buffered with carbonate/CO_2 so there is no requirement for a CO_2 environment for growth. A variety of media have been formulated for insect cell growth. The most commonly employed are Grace's antherea medium (GRACE 1962), Hink's TNM-FH modification of Grace's medium (HINK 1979), TC100 (GARDINER and STOCKDALE 1975), and IPL41 (WEISS et al. 1981), all of which require some serum supplementation. Several companies have developed effective serum-free media for insect cells (e.g., EX-CELL 400, JRH Biosciences, Lexana, Kan.; SF900, GIBCO/BRL, Grand Island, N.Y.) that are rapidly replacing the serum-supplemented formulations. However, these serum-free formulations may not support the growth of all lepidopteran cell lines.

HINK et al. (1991) examined the relative levels of expression of three proteins, beta-galactosidase, human plasminogen, and respiratory syncytial virus gp50T, in 23 lepidopteran cell lines each cultured according to original descriptions. The analysis demonstrated that no individual cell line could be expected to produce optimal levels of all three proteins, and optimal expression of individual proteins for production purposes may require examination of several lepidopteran cell lines.

The expression of recombinant proteins by baculovirus-infected insect cells is efficient enough that 1- to 5-l batch cultures usually suffice for most laboratory scale-up purposes. Roller bottles, air-sparged suspension cultures, and air-lift bioreactors all have proven successful for these purposes (HINK 1982; WEISS et al. 1988). The most convenient method for preparing 1- to 5-l batch cultures is to utilize spinner or stirrer culture setups (HINK 1982; WEISS et al. 1988). These types of cultures are capable of providing cell densities of up to 4 or 5×10^6 cells/ml. Optimal expression of foreign gene products is attained by inoculating during the log phase of the spinner culture growth at densities of around 1×10^6 cells/ml.

The important parameters for suspension cultures of insect cells are shear stress, dissolved oxygen content, and pH (HINK 1982). In a 1-l (or less) batch culture, the pH need not be monitored, assuming that cells are added to fresh medium upon initiating the culture. However, the addition of oxygen to the culture medium by gentle aeration is required to insure optimal cell growth and virus replication (HINK 1982).

The virus inoculum may be added at multiplicities of 0.1 or 1 plaque-forming units (pfu) per cell to log phase cultures of 1×10^6 cells/ml. At these multiplicities the culture may go through one more doubling before all cells become infected. As a conservative estimate, a 1-l culture can be expected to generate 1–10 mg expressed protein product, although levels as high as 200 mg/l are reported (JEANG et al. 1987b).

The product may be harvested between 48 and 72 h p.i., with some attention given to the rate of degradation of product during the course of the infection. LICARI and BAILEY (1991) suggest that as the infection of cells in a bioreactor progresses less cellular proteolytic activity may be evident and the amount of degraded protein product may actually decrease.

Larger scale cultures for the production of baculovirus-expressed proteins are feasible. Air-lift fermenters of up to 30 or 40 l can be employed with either serum-containing or serum-free media to effect large-scale propagation of cells and production of the baculovirus-expressed gene products (MAIORELLA et al. 1988). The first demonstrated serum-free medium, ISFM (INLOW et al. 1989), was based upon the IPL-41 formulation and was effective for both small-scale (less than 10 l) and large-scale (greater than 20 l) production of human macrophage colony stimulating factor (M-CSF) in baculovirus-infected Sf-9 cells (MAIORELLA et al. 1988). The medium substituted a lipid emulsion containing cod liver oil, cholesterol, alpha-tocopherol acetate, and Tween 80 (Sigma) for the serum component, and incorporated pluronic polyol F-68 (BASF Wyandotte) to reduce shear stress on the cells. A 10000-MW filtrate of TC yeastolate reduces additional high-molecular-weight proteins in the medium. Pluronic F-68 has been demonstrated effective in reducing shear dammage to cultured insect cells in agitated, sparged, and air-lift bioreactors (MURHAMMER and GOOCHEE 1988) and is now considered a routine additive for large-scale cultures.

The ISFM formulation provided similar levels of cell growth and recombinant protein production to the serum supplemented IPL-41 medium. Up to 40 mg/l of recombinant M-CSF was produced following infection at cell densities of 3×10^6 cells/ml with a multiplicity of 1 pfu/cell (MAIORELLA et al. 1988). Additional recombinant proteins produced in this manner included plasminogen activator and ricin toxin (see MAIORELLA et al. 1988). These results established that large-scale serum-free cultivation of insect cells for the production of baculovirus-expressed proteins is entirely feasible and can be cost effective. Tissue culture-tested components for the ISFM media formulation are also commercially available (Sigma Chem. Co., St. Louis, Mo.).

Because the virus infection ultimately causes cell death, production cultures are necessarily batch-type rather than continuous. Sequential batch situations for semicontinuous cycling of cells can be envisioned (HINK 1982), but cycling of virus inoculum is not advised due to the problems associated with continued serial propagation of the virus and the generation and amplification of mutations

(BURAND and SUMMERS 1982; KUMAR and MILLER 1987). Each batch culture should be inoculated with virus preparations derived from a stock inoculum that originated from a well-characterized plaque-purified virus. The virus may be stored as infected cell culture supernatants at 4° C for prolonged periods (several months to a year or more) without significant loss of activity. Cell cultures should be started from frozen stocks and used within a defined number of passages. These strategies maximize consistency, and ultimately productivity, for the batch scale-up process.

Finally, several investigators have suggested the possible use of insect larvae as bioreactors for large-scale production of baculovirus-expressed gene products. The silkworm, *B. mori*, has proven an effective alternative to the BM-N cell line (MAEDA et al. 1985; HORIUCHI et al. 1987; MARUMOTO et al. 1987; MIYAJIMA et al. 1987; MAEDA 1989a; TADA et al. 1988; MORISHITA et al. 1991) for production of baculovirus-expressed proteins. The silkworm host is an example of a completely domesticated animal, incapable of survival outside of the laboratory, and thus provides a biological level of containment. The larvae grow to be quite large, and rearing on semisynthetic diets can be automated (MAEDA 1987). As much as 0.3 ml of hemolymph (blood) can be harvested from a single larva, with levels of 30–190 μg protein product/ml (MAEDA et al. 1885; HORIUCHI et al. 1987). The stability of crude preparations of hemolymph appears to be quite good at 4° C, as IL-3 exhibited no loss in activity after storage for several days (MIYAJIMA et al. 1987). Whether all proteins will be as stable remains to be determined.

Strategies have also been developed for production of proteins in *T. ni* larvae (KURODA et al. 1989; PRICE et al. 1989; MEDIN et al. 1990). PRICE et al. (1989) demonstrated the effectiveness of coinfection with wild-type and recombinant baculoviruses to generate infectious OBs containing virions of both viruses, thus increasing the effectiveness of inoculating insect larvae by feeding.

Purification of product from larval hemolymph apparently presents no unusual problems. Both alpha-interferon and IL-3 could be effectively purified by affinity chromatography. In fact, the only difficulty appears to be associated with harvesting by bleeding individual larvae. Presumably an alternative method would be employed in production level scale-up.

6 Concluding Remarks

The use of baculoviruses for expression of eurkaryotic gene products has gained wide acceptance as an attractive alternative to other eurkaryotic expression systems. This is largely due to the ease of handling of the virus and its host cell cultures, and the relatively high levels of expression that can be effected in laboratory-scale production. In addition, the overwhelming majority of recombinant proteins produced by the baculovirus expression system retain the antigenic and biological properties of the native proteins. At the very least, the baculovirus system will continue to be a useful addition to the expression vector repertoire.

Acknowledgements. I would like to thank all those who graciously contributed submitted manuscripts and documents in press. I also thank Dr. Roger Bretthauer and Dr. Subhash Basu for critical reading of portions of the manuscript, Dr. Don Davidson for helpful discussions, and Jozelle Whitmire and Joan Smith for their expert clerical assistance. The expense of preparing this publication was subsidized in part by U.S. Public Health Service grant #AI-22610.

References

Adams JR, Goodwin RH, Wilcox TA (1977) Electron microscopic investigations on invasion and replication of insect baculoviruses in vivo and in vitro. Biol Cell 28: 261–268

Allaway AP, Vivino AA, Kohn LD, Notkins AL, Prabhakar BS (1990) Characterization of the 70 kDa component of the human ku autoantigen expressed in insect cell nuclei using a recombinant baculovirus vector. Biochem Biophys Res Commun 168: 747–755

Andrews DL, Beames B, Summers MD, Park WD (1988) Characterization of the lipid acyl hydrolase activity of the major potato (*Solanum tuberosum*) tuber protein, patatin, by cloning and abundant expression in a baculovirus vector. Biochem J 252: 199–206

Arif, BM (1984) The entomopoxviruses. Adv Virus Res 29: 195–213

Bailey MJ, McLeod DA, Kang C-Y, Bishop, DHL (1989) Glycosylation is not required for the fusion activity of the G protein of vesicular stomatitis virus in insect cells. Virology 169: 323–331

Barber GN, Clegg JCS, Lloyd G (1990) Expression of the Lassa virus nucelocapsid protein in insect cells infected with a recombinant baculovirus: application to diagnostic assays for Lassa virus infection. J Gen Virol 71: 19–28

Baum JA, Geever R, Giles NH (1987) Expression of qa-1F activator protein: Identification of upstream binding sites in the qa gene cluster and localization of the DNA-binding domain. Mol Cell Biol 7: 1256–1266

Baylis SA, Purifoy DJM, Littler E (1991) High-level expression of the Epstein-Barr virus alkaline deoxyribonuclease using a recombinant baculovirus: application to the diagnosis of nasopharyngeal carcinoma. Virology 181: 390–394

Blissard GW, Rohrmann GF (1990) Baculovirus diversity and molecular biology. Annu Rev Entomol 35: 127–155

Brickey DA, Colbran RJ, Fong Y-L, Soderling TR (1990) Expression and characterization of the alpha-subunit of Ca^{2+}/calmodulin-dependent protein kinase II using the baculovirus expression system. Biochem Biophys Res Commun 173: 578–584

Brown CS, Salimans MMM, Noteborn MHM, Weiland HT (1990) Antigenic parvovirus B19 coat proteins VP1 and VP2 produced in large quantities in a baculovirus expression system. Virus Res 15: 197–212

Brown CS, Van Lent JWM, Vlak JM, Spaan WJM (1991) Assembly of empty capsids by using baculovirus recombinants expressing human parvovirus B19 structural proteins. J Virol: 2702–2706

Bud HM, Kelly DC (1980) An electron microscopic study of partially lysed baculovirus nucleocapsids: the intranucleocapsid packaging of viral DNA. J Ultrastruct Res 73: 361–368

Burand JP, Summers MD (1982) Alteration of *Autographa californica* nuclear polyhedrosis virus DNA upon serial passage in cell culture. Virology 119: 223–229

Burand JP, Summers MD, Smith GE (1980) Transfection with baculovirus DNA. Virology 101: 286–290

Bustos MM, Lucknow VA, Griffing LR, Summers MD, Hall TC (1988) Expression, glycosylation and secretion of phaseolin in a baculovirus system. Plant Mol Biol 10: 475–488

Butters TD, Hughes RC, Vischer P (1980) Steps in the biosynthesis of mosquito cell membrane glycoproteins and the effects of tunicamycin. Biochim Biophys Acta 640: 672–686

Cameron IR, Possee RD, Bishop DHL (1989) Insect cell culture technology in baculovirus expression systems. TIBTECH 7: 66–70

Capone JP, Werstuck G (1990) Synthesis of the herpes simplex virus type 1 *trans*-activator Vmw65 in insect cells using a baculovirus vector. Mol Cell Biochem 94: 45–52

Carbonell LF, Klowden MJ, Miller LK (1985) Baculo- virus-mediated expression of bacterial genes in dipteran and mammalian cells. J Virol 56: 153–160

Carbonell LF, Hodge MR, Tomalski MD, Miller LK (1988) Synthesis of a gene coding for an insect-specific scorpion neurotoxin and attempts to express it using baculovirus vectors. Gene 73: 409–418

Carstens EB, Tjia ST, Doerfler W (1979) Infection of *Spodoptera frugiperda* cells with *Autographa californica* nuclear polyhedrosis virus. Virology 99: 386–398

Carstens EB, Tjia ST, Doerfler W (1980) Infectious DNA from *Autographa californica* nuclear polydedrosis virus. Virology 101: 311–314

Chang, LMS, Rafter E, Rusquet-Valerius R, Peterson RC, White ST, Bollum FJ (1988) Expression and processing of recombinant human terminal transferase in the baculovirus system. J Biol Chem 263: 12509–12513

Chen JH (1990) Cloning, sequencing, and expression of mouse c-*ets*-1 cDNA in baculovirus expression system. Oncogene Res 5: 277–285

Chiba J, Ohba H, Matsuura Y, Watanabe Y, Katayama T Kikuchi S, Saito I, Miyamura T (1991) Serodiagnosis of hepatitis C virus (HCV) infection with an HCV core protein molecularly expressed by a recombinant baculovirus. Proc Natl Acad Sci USA 88: 4641–4645

Chiou CJ, Wu MC (1990) Expression of human granulocyte-macrophage colony-stimulating factor gene in insect cells by a baculovirus vector. FEBS Lett 259: 249–253

Chisholm GE, Henner DJ (1988) Multiple early transcripts and splicing of the *Autographa californica* nuclear polyhedrosis virus IE-1 gene. J Virol 62: 3193–3200

Cochran MA, Ericson BL, Knell JD, Smith GE (1987) Use of baculovirus recombinants as a general method for the productiòn of subunit vaccines. In: Chanock R et al. (eds) Vaccines 1987. Cold Spring Harbor Laboratories, Cold Spring Harbor, pp 384–388

Cohen J Charpilienne A, Chilmonczyk S, Estes MK (1989) Nucleotide sequence of bovine rotavirus gene 1 and expression of the gene product in baculovirus. Virology 171: 131–140

Corsaro BG, Fraser MJ (1987a) Characterization of genotypic and phenotypic variation in plaque-purified strains of HzSNPV Elkar isolate. Intervirology 28: 185–198.

Corsaro BG, Fraser MJ (1987b) Characterization of cloned populations of the *Heliothis zea* cell line IPLB-HZ 1075. In Vitro Cell Dev Biol 23: 855–862

Corsaro BG, Fraser MJ (1988) Transfection of Lepidopteran insect cells with baculovirus DNA. J Tissue Cult Methods 12: 7–11

Corsaro BG, DiRenzo J, Fraser JJ (1988) Transfection of cloned *Heliothis zea* cell lines with the DNA genome of the *Heliothis zea* nuclear polyhedrosis virus. J Virol Methods 25: 283–292

Davidson DJ, Castellino FJ (1991a) Asparagine-linked oligosaccharide processing in Lepidopteran insect cells. Temporal dependence of the nature of the oligosaccharides assembled on asparagine-289 of recombinant human plasminogen produced in baculovirus vector-infected *Spodoptera frugiperda* (IPLB-SF-21AE) cells. Biochemistry 30: 6167–6174

Davidson DJ, Castellino FJ (1991b) Structures of the asparagine-289-linked oligosaccharides assembled on recombinant human plasminogen expressed in a *Mamestra brassicae* cell line (IZD-MB0503). Biochemistry 30: 6689–6696

Davidson DJ, Fraser MJ, Castellino FJ (1990) Oligosaccharide processing in the expression of human plasminogen cDNA by Lepidopteran insect (*Spodoptera frugiperda*) cells. Biochemistry 29: 5584–5590

Davidson DJ, Bretthauer RK, Castellino FJ (1991) α-Mannosidase-catalyzed trimming of high-mannose glycans in noninfected and baculovirus-infected *Spodoptera frugiperda* cells (IPLB-SF-21AE). A possible contributing regulatory mechanism for assembly of complex-type oligosaccharides in infected cells. Biochemistry 30: 9811–9815

Delchambre M, Gheysen D, Thines D Thiriart C, Jacobs E, Verdin E, Horth M, Burny A, Bex F (1989). The CAG precursor of simian immunodeficiency virus assembles into virus-like particles. EMBO J 8: 2653–2660

Desprès P, Girard M, Bouloy M (1991) Characterization of yellow fever virus proteins E and NS1 expressed in Vero and *Spodoptera frugiperda* cells. J Gen Virol 72: 1331–1342

Devash Y, Reagan K, Wood D, Turner J, Parrington M, Kang CY (1990) Antibodies against AIDS proteins. Nature 345: 581

DiNocera P, Dawid I (1983) Transient expression of genes introduced into cultured cells of Drosophila. Proc Natl Acad Sci USA 80: 7095–7098

Doerfler W, Bohm P (eds) (1986) The molecular biology of baculoviruses. Curr Top Microbiol Immunol 131. Springer, Berlin Heidelberg New York

Domingo DL, Trowbridge IS (1988) Characterization of the human transferrin receptor produced in a baculovirus expression system. J Biol Chem 263: 13386–13392

Ellis L, Levitan A, Cobb MH, Ramos P (1988) Efficient expression in insect cells of a soluble, active human insulin receptor protein-tyrosine kinase domain by use of a baculovirus vector. J Virol 62: 1634–1639

Emery VC, Bishop DHL (1987) The development of multiple expression vectors for high level synthesis of eurkaryotic proteins: expression of LCMV-N and AcNPV polyhedrin protein by a recombinant baculovirus. Protein Eng 1: 359–366

Estes MK, Crawford SE, Penaranda ME, Petrie B, Burns J, Chan W, Ericson B, Smith G, Summers M (1987) Synthesis and immunogenicity of the rotavirus major capsid antigen using a baculovirus expression system. J Virol 61: 1488–1494

Faulkner P (1981) Baculovirus. Pathogenesis of invertebrate microbial diseases. In: Davidson EW (ed) Allanheld, Osmum, Totowa, pp 1–37

Felleisen R, Beck E, Usmany M, Vlak J, Klinkert M-Q (1990) Cloning and expression of *Schistosoma mansoni* protein Sm32 in a baculovirus vector. Mol Biochem Parasitol 43: 289–292

Fitzpatrick PF, Chlumsky LJ, Daubner SC, O'Malley KL (1990). Expression of rat tyrosine hydroxylase in insect tissue culture cells and purification and characterization of the cloned enzyme. J Biol Chem 265: 2042–2047

Forstova J, Krauzewicz N, Griffin BE (1989) Expression of biologically active middle T antigen of polyoma virus from recombinant beculoviruses. Nucleic Acids Res 17(4): 1427–1443

Fraser MJ (1986a) Ultrastructural observation of virion maturation in *Autographa californica* nuclear polyhedrosis virus-infected *Spodoptera frugiperda* cell cultures. J Ultrastruct Mol Struct Res 95: 189–195

Fraser MJ (1986b) Transposon mediated mutagenesis of baculoviruses: transposon shuttling and implication for speciation. Ann Entomol Soc Am 79: 773–783

Fraser MJ, Hink WF (1982) Comparative sensitivity of several plaque assays techniques employing TN-368 and IPLB-SF21AE insect cell lines for plaque variants of *Galleria mellonella* nuclear polyhedrosis virus. J Invertebr Pathol 40: 89–97

Fraser MJ, Smith GE, Summers MD (1983) Acquisition of host cell DNa sequences by baculoviruses: relationship between host DNA insertions and FP mutants of *Autographa californica* and *Galleria mellonella* nuclear polyhedrosis viruses. J Virol 47: 287–300

French TJ, Inumaru S, Roy P (1989) Expression of two related nonstructural proteins of bluetongue virus (BTV) type 10 in insect cells by a recombinant baculovirus: production of polyclonal ascitic fluid and characterization of the gene product in BTV-infected BHK cells. J Virol: 3270–3278

Fung M-C, Chiu KYM, Weber T, Chang T-W, Chang NT (1988) Detection and purification of a recombinant human B lymphotropic virus (HMV-6) in the baculovirus expression system by limiting dilution and DNA dot-blot hybridization. J Virol Methods 19: 33–42

Funk CD, Gunne H, Steiner H, Izumi T, Samuelsson B (1989) Native and mutant 5-lipoxygenase expression in a baculovirus/insect cell system. Proc Natl Acad Sci USA 86: 2592–2596

Furlong AM, Thomsen DR, Marotti KR, Post LE, Sharma SK (1988) Active human tissue plasminogen activator secreted from insect cells using a baculovirus vector. Biotechnol Appl Biochem 10: 454–464

Gardiner GR, Stockdale H (1975) Two tissue culture media for production of lepidopteran cells and nuclear polyhedrosis viruses. J Invertebr Pathol 25: 363–370

George ST, Arbabian MA, Ruoho AE, Kiely J, Malbon CC (1989) High-efficiency expression of mammalian β-adrenergic receptors in baculovirus-infected insect cells. Biochem Biophys Res Commun 163: 1265–1269

Germann UA, Willingham MC, Pastan I, Gottesman MM (1990) Expression of the human multidrug transporter in insect cells by a recombinant baculovirus. Biochemistry 29: 2295–2303

Giese N, May-Siroff M, LaRochelle WJ, Coelingh KVW, Aaronson SA (1989) Expression and purification of biologically active v-*sis*/platelet-derived growth factor B protein by using a baculovirus vector system. J Virol 63: 3080–3086

Gheysen D, Jacobs E, de Foresta F, Thiriart C, Francotte M, Thines D, de Wilde M (1989) Assembly and release of HIV-1 precursor $Pr55^{gag}$ virus-like particles from recombinant baculovirus-infected insect cells. Cell 59: 103–112

Ginns EI, Rehavi M, Martin BM, Weller M, O'Malley KL, LaMarca ME, McAllister CG, Paul SM (1988). Expression of human tyrosine hydroxylase cDNA in invertebrate cells using a baculovirus vector. J Biol Chem 263: 7406–7410

Goswami BB, Glazer RI (1991) A simplified method for the production of recombinant baculovirus. BioTechniques 10: 626–630
Grabowski GA, White WR, Grace ME (1989) Expression of functional human acid β-glucosidase in COS-1 and *Spodoptera frugiperda* cells. Enzyme 41: 131–142
Grace TDC (1962) Establishment of four strains of cells from insect tissues grown in vitro. Nature 95: 788–789
Graham FL, Van Der Eb AJ (1973) A new technique for the assay of infectivity of human adenovirus 5 DNA. Virology 52: 456–467
Graham F, Bacchetti S, McKinnon R (1980) Transformation of mammalian cells with DNA using the calcium techniques. In: Basegra R, Croce C, Rovera G (eds) Introduction of macromolecules into viable mammalian cells. Liss New York, pp 3–25 (Wistar Symposium Series, Vol 1)
Granados RR, Federici BA (eds) (1986) The biology of baculoviruses. CRC, Boca Raton
Greenfield C, Patel G, Clark S, Jones N, Waterfield MD (1988) Expression of the human EGF receptor with ligand-stimulatable kinase activity in insect cells using a baculovirus vector. EMBO J 7: 139–146
Greenfield C, Hiles I, Waterfield MD, Federwisch M, Wollmer A, Blundell TL, McDonald N (1989) Epidermal growth factor binding induces a conformational change in the external domain of its receptor. EMBO J 8: 4115–4123
Grossman SR, Mora R, Laimins LA (1989) Intracellular localization and DNA-binding properties of human papillomavirus type 18 E6 protein expressed with a baculovirus vector. J Virol: 366–374
Guarino LA, Summers MD (1986) Functional mapping of a trans-activating gene required for expression of a baculovirus delayed-early gene. J Virol 57: 563–571
Guarino LA, Summers MD (1987) Nucleotide sequence and temporal expression of a baculovirus regulatory gene. J Virol 61: 2091–2099
Gunne H, Hellers H, Steiner H (1990) Structure of preproattacin and its processing in insect cells infected with a recombinant baculovirus. Eur J Biochem 187: 699–703
Hammock BD, Bonning BC, Possee RD, Hanzlik TN, Maeda S (1990) Expression and effects of the juvenile hormone esterase in a baculovirus vector. Nature 344: 458–461
Han JH, Law SW, Keller PM, Kniskern PJ, Silberklang M, Tung J-S, Gasic TB, Gasic GJ, Friedman PA, Ellis RW (1989) Cloning and expression of cDNA encoding antistasin, a leech-derived protein having anti-coagulant and anti-metastatic properties. Gene 75: 47–57
Harmon SA, Johnston JM, Ziegelhoffer T, Richards OC, Summers DF, Ehrenfeld E (1988) Expression of hepatitis A virus capsid sequences in insect cells. Virus Res 10: 273–280
Hasemann CA, Capra JD (1990) High-level production of a functional immunoglobulin heterodimer in a baculovirus expression system. Proc Natl Acad Sci USA 87: 3942–3946
Hasnain SE, Nakhai B (1990) Expression of the gene encoding firefly luciferase in insect cells using a baculovirus vector. Gene 91: 135–138
Hauser C, Fusswinkel H, Li J, Oellig C, Kunze R, Muller-Neumann M, Heinlein M, Starlinger P, Doerfler W (1988) Overproduction of the protein encoded by maize transposable element Ac in insect cells by a baculovirus vector
Herrera R, Lebwohl D, de Herreros AG, Kallen RG, Rosen OM (1988) Synthesis, purification, and characterization of the cytoplasmic domain of the human insulin receptor using a baculovirus expression system. J Biol Chem 263: 5560–5568
Hill-Perkins MS, Possee RD (1990) A baculovirus expression vector derived from the basic protein promoter of *Autographa californica* nuclear polyhedrosis virus. J Gen Virol 71: 971–976
Hink WF (1970) Established insect cell line from the Cabbage Looper, *Trichoplusia ni.* Nature 226: 466–467
Hink WF (1979) Cell lines from invertebrates. Methods Enzymol 58: 450–466
Hink WF (1982) Production of *Autographa californica* nuclear polyhedrosis virus in cells from large-scale suspension cultures. In: Kurstak E (ed) Microbial and viral pesticides. Dekker, New York, pp 493–505
Hink WF, Thomsen DR, Davidson DJ, Meyer AL, Castellino FJ (1991) Expression of three recombinant proteins using baculovirus vectors in 23 insect cell lines. Biotechnol Prog 7: 9–14.
Hooft van Iddekinge BJL, Smith GE, Summers MD (1983) Nucleotide sequence of the polyhedrin gene of *Autographa californica* nuclear polyhedrosis virus. Virology 131: 561–565
Horiuchi T, Marumoto Y, Saeki Y, Sato Y, Kurusawa M, Kondo A, Maeda S (1987) High-level expression of the human alpha interferon gene through the use of an improved baculovirus vector in the silkworm, *Bombyx mori.* Agric Biol Chem 51: 1537–1580
Howard SC, Ayres MD, Possee RD (1986) Mapping the 5′ and 3′ ends of *Autographa californica* nuclear polyhedrosis virus polyhedrin mRNA. Virus Res 5: 109–119

Hu SL, Kosowski SG, Schaaf KF (1987) Expression of envelope glycoproteins of human immunodeficiency virus by an insect virus vector. J Virol 61: 3617–3620
Hu Y-W, Kang CY (1991) Enzyme activities in four different forms of human immunodeficiency virus 1 *pol* gene products. Proc Natl Acad Sci USA 88: 4596–4600
Hughes KM (1972) Fine structure and development of two polyhedrosis viruses. J Invertebr Pathol 19: 198–207
Iacono-Connors LC, Schmaljohn CS, Dalrymple JM (1990) Expression of the *Bacillus anthracis* protective antigen gene by baculovirus and vaccinia virus recombinants. Infect Immun 58: 366–372.
Iatrou K, Ito K, Witkiewicz H (1985) Polyhedrin gene of *Bombyx mori* nuclear polyhedrosis virus. J Virol 54: 436–445
Iatrou K, Meidinger RG, Goldsmith MR (1989) Recombinant baculoviruses as vectors for identifying proteins encoded by intron-containing members of complex multigene families. Proc Natl Acad Sci USA 86: 9129–9133
Inlow D, Shauger A, Maiorella B (1989) Insect cell culture and baculovirus propatation in protein-free medium. J Tissue Cult Methods 12: 13–16
Inumaru S, Roy P (1987) Production and characterization of the neutralization antigen VP2 of Bluetongue virus serotype 10 using a baculovirus expression vector. Virology 157: 472–479
Inumaru S, Ghiasi H, Roy P (1987) Expression of Blue- tongue virus group-specific antigen VP3 in insect cells baculovirus vector: its use for the detection of Bluetongue virus antibodies. J Gen Virol 68: 1627–1635
Jansen JJM, Mulder WR, De Clauwè GLJ, Vlak JM, DeGrip WJ (1991) In vitro expression of bovine opsin using recombinant baculovirus: the role of glutamic acid (134) in opsin biosynthesis and glycosylation. Biochim Biophys Acta 1089: 68–76
Jarvis DL, Summers MD (1989) Glycosylation and secretion of human tissue plasminogen activator in recombinant baculovirus-infected insect cells. Mol Cell Biol 9: 214–223
Jeang KT, Holmgren-Konig M, Khoury G (1987a). A baculovirus vector can express intron-containing genes. J Virol 61: 1761–1764
Jeang KT, Giam CZ, Nierenberg M, Khoury G (1987b) Abundant synthesis of functional HTLV-1 p40x protein in eukaryotic cells by using a baculovirus expression vectors. J Virol 61: 708–713
Jeang KT, Shank PR, Rabson AB, Kumar A (1988a) Synthesis of functional human immunodeficiency virus *tat* protein in baculovirus as determined by a cell-cell fusion assay. J Virol 62: 3874–3878
Jeang KT, Shank PR, Kumar A (1988b) Transcriptional activation of homologous viral long terminal repeats by the human immunodeficiency virus type 1 or the human T-cell leukemia virus type I tat proteins occurs in the absence of de novo protein synthesis. Proc Natl Acad Sci USA 85: 8291–8295
Johnson PW, Attia J, Richardson CD, Roder JC, Dunn RJ (1989) Synthesis of soluble myelin-associated glycoprotein in insect and mammalian cells. Gene 77: 287–296
Kajigaya S, Fujii H, Field A, Anderson S, Rosenfeld S, Anderson LJ, Shimada T, Young NS (1991) Self-assembled B19 parvovirus capsids, produced in a baculovirus system, are antigenically and immunogenically similar to native virions. Proc Natl Acad Sci USA 88: 4646–4650
Kang CY (1988) Baculovirus vectors for expression of foreign genes. Adv Virus Res 35: 177–192
Kang CY, Bishop DHL, Seo JS, Matsuma Y, Choe M (1987) Secretion of particles of hepatitis B surface antigen from insect cells using a baculovirus vector. J Gen Virol 68: 2607–2613
Kawai S, Nishiyawa M (1984) New procedure for DNA transfection with polycation and dimethyl sulfoxide. Mol Cell Biol 4: 1172–1174
Kelly DC (1982) Baculovirus replication. J Gen Virol 63: 1–13
Kelly DC, Lescott T (1981) Baculovirus replication: Protein synthesis in *Spodoptera frugiperda* cells infected with *Trichoplusia ni* nuclear polyhedrosis virus. Microbiologica 4: 35–57
Kitts PA, Ayres MD, Possee RD (1990) Linearization of baculovirus DNA enhances the recovery of recombinant virus expression vectors. Nucleic Acids Res 18: 5667–5672
Klaiber K, Williams N, Roberts TM, Papazian DM, Jan LY, Miller C (1990) Functional expression of *Shaker* K^+ channels in a baculovirus-infected inseet cell line. Neuron 5: 221–226
Klockmann U, Deppert W (1983) Acylation: a new post-translational modification specific for plasma membrane-associated simian virus 40 large T-antigen. FEBS Lett 151: 257–259
Knudson DL (1979) Plaque assay of baculoviruses employing an agarose-nutrient overlay. Intervirology 11: 40–46
Knudson DL, Harrap KA (1976) Replication of a nuclear polyhedrosis virus in a continuous cell culture of *Spodoptera frugiperda*: microscopy study of the sequence of events of the virus infection. J Virol 17: 254–268

Koener JF, Leong JC (1990) Expression of the glycoprotein gene from a fish rhabdovirus by using baculovirus vectors. J Virol: 428–430
Krishna S, Blacklaws BA, Overton HA, Bishop DHL, Nash AA (1989) Expression of glycoprotein D of herpes simplex virus type 1 in a rocombiant baculovirus: protective responses and T cell recognition of the recombinant-infected cell extracts. J Gen Virol 70: 1805–1814
Kristie TM, LeBowitz JH, Sharp PA (1989) The octamer-binding proteins form multi-protein-DNA complexes with the HSV αTIF regulatory protein. EMBO J 8: 4229–4238
Kumar S, Miller LK (1987) Effects of serial passage of *Autographa californica* nuclear polyhedrosis virus in cell culture. Virus Res 7: 335–349
Kuroda K, Hauser C, Rott R, Klenk HD, Doerfler W (1986) Expression of the influenza virus haemagglutinin in insect cells by a baculovirus vector. EMBO J 5: 1359–1365
Kuroda K, Hauser C, Rott R, Klenk HD, Doerfler W (1987) Biologically active influenza virus haemagglutinin expressed in insect cells by a baculovirus vector. In: Maramorosch K (ed) Biotechnology in invertebrate pathology land and cell culture. Academic, New York, pp 235–249
Kuroda K, Gröner A, Frese K, Drenckhahn D, Hauser C, Rott R, Doerfler W, Klenk H-D (1989) Synthesis of biologically active influenza virus hemagglutinin in insect larvae. J Virol 63: 1677–1685
Kuroda K, Geyer H, Geyer R, Doerfler W, Klenk H-D (1990) The oligosaccharides of influenza virus hemagglutinin expressed in insect cells by a baculovirus vector. Virology 174: 418–429
Kuzio J, Rohel DZ, Curry CJ, Krebs A, Carstens EB, Faulkner P (1984) Nucleotide sequence of the p10 polypeptide gene of *Autographa californica* nuclear polyhedrosis virus. Virology 139: 414–418
Lanford RE (1988) Expression of simian virus 40 T antigen in insect cells using a baculovirus expression vector. Virology 167: 72–81
Lanford RE, Kennedy RC, Dreesman GR, Eichberg JW, Notvall L, Luckow VA, Summers MD (1988) Expression of hepatitis B virus surface and core antigens using a baculovirus expression vector. In: Zuckerman AJ (ed) Viral hepatitis and liver disease. Liss, New York, pp 372–378
Lebacq-Verheyden AM, Kasprzyk PG, Raum MG, Van Wyke Coelingh K, Lebacq JA, Battey JF (1988) Posttranslational processing of endogenous and of baculovirus-expressed human gastrin-releasing peptide precursor. Mol Cell Biol 8: 3129–3135
Leisy D, Rohrmann G, Beaudreau G (1986a) The nucleotide sequence of the polyhedrin gene region from the multicapsid baculovirus of *Orgyia pseudotsugata*. Virology 153: 280–288
Leisy D, Rohrmann G, Nesson M, Beaudreau G (1986b) Nucleotide sequencing and transcriptional mapping of the *Orgyia pseudotsugata* multicapsid nuclear polyhedrosis virus p10 gene. Virology 153: 157–167
Licari P, Bailey JE (1991) Factors influencing recombinant proteins yields in an insect cell-baculovirus expression system: multiplicity of infection and intracellular protein degradation. Biotechnol Bioeng 37: 238–246
Lipes MA, Napolitano M, Jeang K-T, Chang NT, Leonard WJ. Identification, cloning, and characterization of an immune activation gene. Proc Natl Acad Sci USA 85: 9704–9708
Livingston C, Jones I (1989) Baculovirus expression vectors with single strand capability. Nucleic Acids Res 17: 2366
Lowe PN, Page MJ, Bradley S, Rhodes S, Sydenham M, Paterson H, Skinner RH (1991) Characterization of recombinant human Kirsten-*ras* (4B) p21 produced at high levels in *Escherichia coli* and insect baculovirus expression systems. J Biol Chem 266: 1672–1678.
Luckow VA (1991) Cloning and expression of heterologous genes in insect cells with baculovirus vectors. In: Prokop A, Bajpai RK, Ho CS (Eds), Recombinant DNA Technology and Applications. McGraw-Hill, New York, pp 97–151
Luckow VA, Summers MD (1988a) Trends in the development of baculovirus expression vectors. Biotechnology 6: 47–55
Luckow VA, Summers MD (1988b) Signals important for high-level expression of foreign genes in *Autographa californica* nuclear polyhedrosis virus expression vectors. Virology 167: 56–71
Luckow VA, Summers MD (1989) High level expression of nonfused foreign genes with *Autographa* californica nuclear polyhedrosis virus expression vectors. Virology 170: 31–39
Mackow ER, Barnett JW, Chan H, Greenberg HB (1989) The rhesus rotavirus outer capsid protein VP4 functions as a hemagglutinin and is antigenically conserved when expressed by a baculovirus recombinant. J Virol, pp 1661–1668
Mackow ER, Vo PT, Broome R, Bass D, Greenberg HB (1990) Immunization with baculovirus-expressed VP4 protein passively protects against simian and murine rotavirus challenge. J. Virol, pp 1698–1703
Madisen L, Travis B, Hu SL, Purchio AF (1987) Expression of the human immunodeficiency virus gag gene in insect cells. Virology 158: 248–250

Maeda S (1987) Expression of human interferon alpha in silkworms with a baculovirus vector. In: Maramorosch K (ed) Biotechnology in invertebrate pathology and cell culture. Academic, New York, pp 222–232
Maeda S (1989a) Expression of foreign genes in insects using baculovirus vectors. Annu Rev Entomol 34: 351–372
Maeda S (1989b) Increased insecticidal effect by a recombinant baculovirus carrying a synthetic diuretic hormone gene. Biochem Biophys Res Commun 165: 1177–1183
Maeda S, Kawai T, Obinata M, Fujiwara H, Horiuchi T, Saeki Y, Sato Y, Furasawa M (1985) Production of human alpha interferon in silkworms using a baculovirus vector. Nature 315: 592–594
Maiorella B, Inlow D, Shauger A, Harano D (1988) Large-scale insect cell-culture for recombinant protein production. Biotechnology 6: 1406–1410
Makino Y, Tadano M, Anzai T, Ma SP, Yasuda S, Fukunaga T (1989) Detection of dengue 4 virus core protein in the nucleus. II. Antibody against dengue 4 core protein produced by a recombinant baculovirus reacts with the antigen in the nucleus. J Gen Virol 70: 1417–1425
Mann SG, King LA (1989) Efficient transfection of insect cells with baculovirus DNA using electroporation. J Gen Virol 70: 3501–3505
Marcy AI, Olivo PD, Challberg MD, Coen DM (1990) Enzymatic activities of overexpressed herpes simplex virus DNA polymerase purified from recombinant baculovirus-infected insect cells. Nucleic Acids Res 18: 1207–1215
Marshall JA, Roy P (1990) High level expression of the two outer capsid proteins of bluetongue virus serotype 10: their relationship with the neutralization of virus infection. Virus Res 15: 189–196
Marshall JA, Fayard B, Roy P (1990) Biophysical studies on the morphology of baculovirus-expressed bluetongue virus tubules. J Gen Virol 71: 1839–1844
Martens JWM, Honée G, Zuidema D, Van Lent JWM, Visser B, Vlak JM (1990) Insecticidal activity of a bacterial crystal protein expressed by a recombinant baculovirus in insect cells. Appl Environ Microbiol 56: 2764–2770
Martin BM, Tsuji S, LaMarca ME, Maysak K, Eliason W, Ginns EI (1988) Glycosylation and processing of high levels of active human glucocerebrosidase in invertebrate cells using a baculovirus expression vector. DNA 7: 99–106
Marumoto Y, Sata Y, Fujiwara H, Sakano K, Saeki Y, Agata M, Furusawa M, Maeda S (1987) Hyperproduction of polyhedrin-IGF II fusion protein in silkworm larvae infected with recombinant *Bombyx mori* nuclear polyhedrosis virus. J Gen Virol 68: 2599–2606
Matsuura Y, Possee RD, Bishop DHL (1986) Expression of the s-coded genes of lymphocytic choriomeningitis arenavirus using a baculovirus vector. J Gen Virol 67: 1515–1529
Matsuura Y, Possee RD, Overton HA, Bishop DHL (1987) baculovirus expression vectors: The requirements for high level expression of proteins, including glycoproteins. J Gen Virol 68: 1233–1250
Matthews REF (1982) Classification of nomenclature of viruses. Intervirology 17: 1–199
Medin JA, Hunt L, Gathy K, Evans RK, Coleman MS (1990) Efficient, low-cost protein factories: expression of human adenosine deaminase in baculovirus-infected insect larvae. Proc Natl Acad Sci USA 87: 2760–2764
Miller DW, Miller LK (1982) A virus mutant with an insertion of a copia-like transposable element. Nature 299: 562–564
Miller DW, Safer P, Miller LK (1986) An insect baculovirus host vector system for high-level expression of foreign genes. Gen Eng 8: 277
Miller LK (1988) Baculoviruses for foreign gene expression in insect cells In: Rodriquez RL, Denhardt DT (eds) Vectors: a survey of molecular cloning vectors and their uses. Butterworth, Boston, pp 457–465
Miller LK, Miller DW, Adang MJ (1983a) An insect virus for genetic engineering: developing baculovirus polyhedrin substitution vectors. In: Lurgin PF, Kleinhofs A (eds) Genetic engineering in eukaryotes. Plenum, New York, pp 89–98
Miller LK, Trimarchi RE, Browne D, Pennock GD (1983b) A temperature-sensitive mutant of the baculovirus *Autographa californica* nuclear polyhedrosis virus defective in an early function required for further gene expression. Virology 126: 376–380
Mills HR, Jones IM (1990) Expression and purification of p24, the core protein of HIV, using a baculovirus-insect cell expression system. AIDS 4: 1125–1131
Miyagawa K, Sakamoto H, Yoshida T, Yamashita Y, Mitsui Y, Furusawa M, Maeda S, Takaku F, Sugimura T, Terada M (1988) *hst-1* transforming protein: expression in silkworm cells and characterization as a novel heparin-binding growth factor. Oncogene 3: 383–389

Miyajima A, Schreurs J, Otsu K, Kondo A, Arai K, Maeda S (1987) Use of the silkworm, *Bombyx mori*, and an insect baculovirus vector for high level expression and sectretion of biologically active mouse interleukin-3. Gene 58: 273–281

Miyamoto C, Smith GE, Farrell-Towt J, Chizzonite R, Summers MD, Ju G (1985) Production of human c-*myc* protein in insect cells infected with a baculovirus expression vector. Mol Cell Biol 5: 2860–2865

Morikawa S, Booth TF, Bishop DHL (1991) Analyses of the requirements for the synthesis of virus-like particles by feline immunodeficiency virus *gag* using baculovirus vectors. Virology 183: 288–297

Morishita K, Sakano K, Takeda K, Maeda S (1991) Characterization of v-*sis* protein expressed in silkworm larvae using the *Bombyx mori* nuclear polyhedrosis virus vector. J Biochem 109: 36–44

Murhammer DW, Goochee CF (1988) Scaleup of insect cell cultures: protective effects of pluronic F-68. Biotechnology 6: 1411–1418

Murphy CI, Weiner B, Bikel I, Piwnica-Worms H, Bradley MK, Livingston DM (1988) Purification and functional properties of simian virus 40 large and small T antigens overproduced in insect cells. J Virol 62: 2951–2959

Nagy É, Derbyshire JB, Dobos P, Krell PJ (1990) Cloning and expression of NDV hemagglutinin-neuraminidase cDNA in a baculovirus expression vector system. Virology 176–438

Nakhai B, Pal R, Sridhar P, Talwar GP, Hasnain SE (1991) The α subunit of human chorionic gonadotropin hormone synthesized in insect cells using a baculovirus vector is biologically active. FEBS Lett 283: 104–108

Neufeld KL, Richards OC, Ehrenfeld E (1991) Expression and characterization of poliovirus proteins $3B^{VPg}$, $3C^{prl}$ in recombinant baculovirus-infected *Spodotera frugiperda* cells. Virus Res 19: 173–188

Niikura M, Matsuura Y, Hattori M, Onuma M, Mikami T (1991a) Characterization of haemogglutinin-neuraminidase glycoprotein of Newcastle disease virus expressed by a recombinant baculovirus. Virus Res 20: 31–43

Niikura M, Matsuura Y, Hattori M, Onuma M, Mikami T (1991b) Expression of the A antigen (gp57-65) of Marek's disease virus by a recombinant baculovirus. J Gen Virol 72: 1099–1104

Nyunoya H, Akagi T, Ogura T, Maeda S, Shimotohno K (1988) Evidence for phosphorylation of *trans*-activator p40* of human T-cell leukemia virus type 1 produced in insect cells with a baculovirus expression vector. Virology 167: 538–544

Oker-Blom C, Summers MD (1989) Expression of Sindbis virus 26S cDNA in *Spodoptera frugiperda* (Sf9) Cells, using a baculovirus expression vector. J Virol 63: 1256–1264

Oldfield S, Adachi A, Urakawa T, Hirasawa T, Roy P (1990) Purification and characterization of the major group-specific core antigen VP7 of bluetongue virus synthesized by a recombinant baculovirus. J Gen Virol 71: 2649–2656

Olivo PD, Nelson NJ, Challberg MD (1988) Herpes simplex virus DNA replication: the UL9 gene encodes an origin-binding protein. Proc Natl Acad Sci USA 85: 5414–5418

Olivo PD, Nelson NJ, Challberg MD (1989) Herpes simplex virus type 1 gene products required for DNA replication: identification and overexpression. J Virol 63: 196–204

Ollo R, Maniatis T (1987) Drosophila *Kruppel* gene produced in a baculovirus expression system is a nuclear phosphoprotein that binds to DNA. Proc Natl Acad Sci USA 84: 5700–5704

Ooi BG, Rankin C, Miller LK (1989) Downstream sequences augment transcription from the essential initiation site of a baculovirus polyhedrin gene. J Mol Biol 210: 721–736

O'Reilly DR, Miller LK (1988) Expression and complex formation of simian virus 40 large T antigen and mouse p53 in insect cells. J Virol 62: 3109–3119

O'Reilly DR, Miller LK (1990) Regulation of expression of a baculovirus ecdysteroid UDP-glucosyltransferase gene. J Virol, pp 1321–1328

Overton HA, Fujii Y, Price IR, Jones IM (1989) The protease and *gag* gene products of the human immunodeficiency virus: authentic cleavage and post-translational modication in an insect cell expression system. Virology 170: 107–116

Page MJ (1989) p36C: an improved baculovirus expression vector for producing high levels of mature recombinant proteins. Nucleic Acids Res 17: 454

Page MJ, Hallk A, Rhodes S, Skinner RH, Murphy V, Sydenham M, Lowe PN (1989) Expression and characterization of the Ha-*ras* p21 protein produced at high levels in the insect/baculovirus system. J Biol Chem 264: 19147–19154

Parekh RB, Dwek RA, Edge CJ, Rademacher TW (1989) N-glycosylation and the production of recombinant glycoproteins. TIBTECH 7: 117–121

Parker EM, Kameyama K, Higashijima T, Ross EM (1991) Reconstitutively active G porotein-coupled receptors purified from baculovirus-infected insect cells. J Biol Chem 266: 519–527

Patel G, Stabel S (1989) Expression of a functional protein kinase $C - \gamma$ using a baculovirus vector: purification and characterisation of a single protein kinase C iso-enzyme. Cell Sign 1: 227–240
Paul JI, Tavaré J, Denton RM, Steinert DF (1990) Baculovirus-directed epression of the human insulin receptor and an insulin-binding ectodomain. J Biol Chem 265: 13074–13083
Payne CC, Mertens PPC (1983) Cytoplasmic polyhedrosis viruses. In: Joklik WK (ed) The Reoviridae. Academic, New York, pp 425–504
Pen J, Welling GW, Welling-Wester S (1989) An efficient procedure for the isolation of recombinant baculovirus. Nucleic Acids Res 17: 451
Pennock GD, Shoemaker C, Miller LK (1984) Strong and regulated expression of *E. coli* beta-galactosidase in insect cells using a baculovirus vector. Mol Cell Biol 4: 399–406
Persing DJ, Varmus HE, Ganem D (1987) The pre-S1 protein of hepatitis B virus is acylated at its amino terminus with myristic acid. J Virol 61: 1672–1677
Possee RD (1986) Cell-surface expression of influenza virus haemagglutinin in insect cells using a baculovirus vector. Virus Res 5: 43–59
Possee RD, Howard SC (1987) Analysis of the polyhedrin gene promoter of the *Autographa californica* nuclear polyhedrosis virus. Nucleic Acids Res 15: 10233–10248
Potter KM, Miller LK (1980a) Transfection of two invertebrate cell lines with the DNA of *Autographa californica* nuclear polyhedrosis virus. J Invertebr Pathol 36: 431–432
Potter KN, Miller LK (1980b) Correlating genetic mutations of a baculovirus with the physical map of the DNA genome. In: Field BN, Jaenisch R (eds) Animal virus genetics. Academic, New York, pp 71–80
Préhaud C, Harris RD, Fulop V, Koh C-L, Wong J, Flamand A, Bishop DHL (1990) Expression, characterization, and purification of a phosphorylated rabies nucleoprotein synthesized in insect cells by baculovirus vectors. Virology 178: 486–497
Price PM, Reichelderfer CF, Johansson BE, Kilbourne ED, Acs G (1989) Complementation of recombinant baculoviruses by coinfection with wild-type virus facilitates production in insect larvae of antigenic proteins of hepatitis B virus and influenza virus. Proc Natl Acad Sci 86: 1453–1456
Quant-Russell R, Pearson MN, Rohrmann GF, Beaudreau G (1987) Characterization of baculovirus p10 synthesis using monoclonal antibodies. Virology 160: 9–19
Quelle FW, Casiake LF, Burkert RE, Wojchowski DM (1989) High-level expression and purification of a recombinant human erythropoietin produced using a baculovirus vector. Blood 74: 652–657
Quilliam LA, Der CJ, Clark R, O'Rourke EC, Zhang K, McCormick F, Bokoch GM (1990) Biochemical characterization of baculovirus-expressed *rap*1A K*rev*-1 and its regulation by GTPase-activating proteins. Mol Cell Biol 10: 2901–2908
Ragona G, Edwards SA, Mercola DA, Adamson ED, Calogero A (1991) The transcriptional factor Egr-1 is synthesized by baculovirus-infected insect cells in an active, DNA-binding form. DNA Cell Biol 10: 61–66
Rankin C, Ooi BG, Miller LK (1988) Eight base pairs encompassing the transcriptional start point are the major determinant for baculovirus polyhedrin gene expression. Gene 70: 39–50
Rasmussen L, Battles JK, Ennis WH, Ngashima K, Gonda MA (1990) Characterization of virus-like particles produced by a recombinant baculovirus containing the *gag* gene of the bovine immunodeficiency-like virus. Virology 178: 435–451
Ray R, Galinski MS, Compans RW (1989) Expression of the fusion glycoprotein of human parainfluenze type 3 virus in insect cells by a recombinant baculovirus and analysis of its immunogenic property. Virus Res 12: 169–180
Reiländer H, Boege F, Vasudevan S, Maul G, Hekman M, Dees C, Hampe W, Helmreich EJM, Michel H (1991) Purification and functional characterization of the human β_2-adrenergic receptor produced in baculovirus-infected insect cells. FEBS Lett 282: 441–444
Rice WC, Lorimar HE, Prives C, Miller LK (1987) Expression of polyoma virus large T antigen by using a baculovirus vector. J Virol 61: 1712–1716
Rodewald HR, Langhorne J, Eichmann K, Kupsch J (1990) Production of murine interleukin-4 and interleukin-5 by recombinant baculovirus. J Immunol Methods 132: 221–226
Rohrmann GF (1986) Polyhedrin structure. J Gen Virol 67: 1499–1513
Roosien J, Belsham GJ, Ryan MD, King AMQ, Vlak JM (1990) Synthesis of foot-and-mouth disease virus capsid proteins in insect cells using baculovirus expression vectors. J Gen Virol 71: 1703–1711
Rose RC, Bonnez W, Strike DG, Reichman RC (1990) Expression of the full-length products of the human papillomavirus type 6b (HPV-6b) and HPV-11 L2 open reading frames by

recombinant baculovirus, and antigenic comparisons with HPV-11 whole virus particles. J Gen Virol 71: 2725–2729
Rota PA, Black RA, De BK, Harmon MW, Kendal AP (1990) Expression of influenza A and B virus nucleoprotein antigens baculovirus. J Gen Virol 71: 1545–1554
Rothman AL, Kurane I, Zhang YM, Lai C-J, Ennis FA (1989) Dengue virus-specific murine T-lymphocyte proliferation: serotype specificity and response to recombinant viral proteins. J Virol 63: 2486–2491
Rusche JR, Lynn DL, Robert-Guroff M, Langlois AJ, Lyerly HK, Carson H, Krohn K, Ranki A, Gallo RC, Bolognesi DP, Putney SD, Matthews TJ (1987) Humoral immune response to the entire human immunodeficiency virus envelope glycoprotein made in insect cells. Proc Natl Acad Sci USA 84: 6924–6928
Sánchez-Martínez D, Pellett PE (1991) Expression of HSV-1 and HSV-2 glycoprotein G in insect cells by using a novel baculovirus expression vector. Virology 182: 229–238
Schmaljohn CS, Sugiyama K, Schmaljohn AL, Bishop DHL (1988) Baculovirus expression of the small genome segment of Hantaan virus and potential use of the expressed nucleocapsid protein as a diagnostic antigen. J Gen Virol 69: 777–786
Schmaljohn CS, Parker MD, Ennis WH, Dalrymple JM, Collett MS, Suzich JA, Schmaljohn AL (1989) Baculovirus expression of the M genome segment of Rift Valley fever virus and examination of antigenic an immunogenic properties of the expressed proteins. Virology 170: 184–192
Schmaljohn CS, Chu Y-K, Schmaljohn AL, Dalrymple jM (1990) Antigenic subunits of Hantaan virus expressed by baculovirus and vaccinia virus recombinants. J Virol 64: 3162–3170
Scully LJ, Kang CY (1988) Production and secretion of hepatitis B surface antigen (HBsAg) and Pre-S2 plus HBsAg using a baculovirus expression vector. In: Zuckerman AJ (ed) Viral hepatitis and liver disease. Liss, New York, pp 365–371
Sekine H, Fuse A, Inaba, Noriyuki, Takamizawa H, Simizu B (1989) Detection of the human papillomavirus 6b E2 gene product in gneital condyloma and laryngeal papilloma tissues. Virology 170: 92–98
Shiu SYW, Morikawa S, Buckley A, Higgs S, Karunakarannair V, Balchere C, Gould EA (1991) 17D yellow fever vaccine virus envelope protein expressed by recombinant baculovirus is antigenically indistinguishable from authentic viral protein. J Gen Virol 72: 1451–1454
Sissom J, Ellis L (1989) Secretion of the extracellular domain of the human insulin receptor from insect cells by use of a baculovirus vector. Biochem J 261: 119–126
Smith GE, Vlak JM, Summers MD (1983a) Physical analysis of *Autographa californica* nuclear polyhedrosis virus transcripts for polyhedrin and 10,000-molecular-weight protein. J Virol 45: 215–225
Smith GE, Fraser MJ, Summers MD (1983b) Molecular engineering of the *Autographa californica* nuclear polyhedrosis virus genome: deletion mutations within the polyhedrin gene. J Virol 46: 584–593
Smith GE, Summers MD, Fraser MJ (1983c) Production of human beta interferon in insect cells infected with a baculovirus expression vector. Mol Cell Biol 3: 2156–2165
Smith GE, Ju G, Ericson BL, Moschera J, Lahm HW, Chizzonite R, Summers MD (1985) Modification and secretion of human interleukin 2 produced in insect cells by a baculovirus expression vector. Proc Natl Acad Sci USA 82: 8404–8408
Spandan DF, Wang HH, Fraser MJ, Lee C (1991) A functional hepatitis B virus X protein produced in insect cells. Virology (in press)
St Angelo C, Smith GE, Summers MD, Krug RM (1982) Two of three influenza viral polymerase proteins expressed by using baculovirus vectors form a complex in insect cells. J Virol 61: 361–365
Steiner H, Polh G, Hellers M, Elhammer A, Hansson L (1988) Human tissue-type plasminogen activator synthesized by using a baculovirus vector in insect cells compared with human plasminogen activator produced in mouse cells. Gene 73: 449–457
Stewart LMD, Hirst M, López F, Merryweather AT, Cayley PJ, Possee RD (1991) Construction of an improved baculovirus insecticides containing an insect-specific toxin tene. Nature 352: 1991
Stoltz DB, Pavan C, Da Cunha AB (1973) Nuclear polyhedrosis virus a possible example of de novo intranuclear membrane morphogenesis. J Gen Virol 19: 145–150
Summers MD, Smith GE (1976) Comparative studies of baculovirus granulins and polyhedrins. Intervirology 6: 168–180
Simmers MD, Smith GE (1987) A manual of methods for baculovirus vectors and insect cell culture procedures. Tex Agricult Exp Stat Bull 1555

Tada A, Fuse A, Sekine H, Simizu B, Kondo A, Maeda S (1988) Expression of the E2 open reading frame of papillomaviruses BPV 1 and HPV6b in silkworm by a baculovirus vector. Virus Res 9: 357–367

Takehara K, Ireland D, Bishop DHL (1988) Co-expression of the hepatitis B surface and core antigens using baculovirus multiple expression vectors. J Gen Virol 69: 2763–2777

Tarentino AL, Maley F (1974) Purification and properties of an endo-Beta-*N*-acetylglucosaminidase from *Streptomyces ginseus*. J Biol Chem 249: 811–817

Thomas CP, Booth TF, Roy P (1990) Synthesis of bluetongue virus encoded phosphoprotein and formation of inclusion bodies by recombinant baculovirus in insect cells: it binds the single-stranded RNA species. J Gen Virol 71: 2073–2083

Thomsen DR, Post LE, Elhammer, Ethammer AP (1990) Structure of *O*-glycosidically linked oligosaccharides synthesized by the insect cell line Sf9. J Cell Biochem 43: 67–79

Tomalski MD, Miller LK (1991) Insect paralysis by baculovirus mediated expression of a mide neurotoxin gene. Nature 352: 82–85

Tratner I, de Togni P, S-C P, Verma IM (1990) Characterization and purification of human *fos* protein generated in insect cells with a baculoviral expression vector. J Virol 64: 499–508

Tweeten KA, Bulla LA Jr, Consigli RA (1980) Characterization of an extremely basic protein derived from granulosis virus nucleocapsids. J Virol 33: 866–876

Urakawa T, Roy P (1988) Bluetongue virus tubules made in insect cells by recombinant baculoviruses: expression of the NS1 gene of bluetongue virus serotype 10. J Virol 62: 3919–3927

Urakawa T, Ferguson M, Minor PD, Cooper J, Sullivan M, Almond JW, Bishop DHL (1989) Synthesis of immunogenic, but non-infectious, poliovirus particles in insect cells by a baculovirus expression vector. J Gen Virol 70: 1453–1463

Van Bokhoven H, Wellink J, Usmany M, Vlak JM, Goldbach R, van Kammen A (1990) Expression of plant virus genes in animal cells: high-level synthesis of cowpea mosaic virus B-RNA-encoded proteins with baculovirus expression vectors. J Gen Virol 71: 2509–2517

Van der Wilk F, van Lent JWM, Vlak JM (1987) Immunogold detection of polyhedrin $p10_1$ and virion antigens in *Autographa californica* nuclear polyhedrosis virus-infected *Spodoptera frugiperda* cells. J Gen Virol 68: 2615–2623

Van Drunen Littel-van den Hurk S, Parker MD, Fitzpatrick DR, Žamb TJ, van den Hurk JV, Campos M, Harland R, Babiuk LA (1991) Expression of bovine herpesvirus 1 glycoprotein gIV by recombinant baculovirus and analysis of its immunogenic properties. J Virol 65: 236–271

Van Wyke Coelingh KLV, Murphy BR, Collins PL, Lebacq-Verheyden AM, Battey JF (1987) Expression of biologically active and antigenically authentic parainfluenza type 3 virus Haemagglutinin–Neuraminidase glycoprotein by a recombinant baculovirus. Virology 160: 465–472

Vaughn JL, Goodwin RH, Tompkins GJ, McCawley P (1977) The establishment of two cell lines from the insect *Spodoptera frugiperda* (Lepidoptera: Noctuidae). In Vitro 13: 213–217

Vialard J, Lalumière, Vernet T, Briedis D, Alkhatib G, Henning D, Levin D, Richardson C (1990) Synthesis of the membrane fusion and hemagglutinin proteins of measles virus, using a novel baculovirus vector containing the β-galactosidase gene. J Virol 64: 37–50

Vik TA, Sweet LJ, Erikson RL (1990) Coinfection of insect cells with recombinant baculovirus expressing $pp60^{v-src}$ results in the activation of a serine-specific protein kinase $pp90^{rsk}$. Proc Natl Acad Sci USA 87: 2685–2689

Villalba M, Wente SR, Russell DS, Ahn J, Reichedderfer CF, Rosen OM (1989) Another version of the human insulin receptor kinase domain: expression, purification, and characterization. Proc Natl Acad Sci USA 86: 7848–7852

Villareal LP, Berg P (1977) Hybridization in situ of SV40 plaques: detection of recombinant SV40 virus carrying specific sequences of non-viral DNA. Science 196: 183–185

Vissavajjhala P, Ross AH (1990) Purification and characterization of the recombinant extracellular domain of human nerve growth factor receptor expressed in a baculovirus system. J Biol Chem 265: 4746–4752

Vlak JM, Keus RJA (1990) Baculovirus expression vector system for production of viral vaccines. In: Mizrahi A (ed) Viral vaccines. Wiley, New York, pp 91–128

Vlak JM, Rohrmann GF (1985) The nature of polyhedrin. In: Maramorosch K, Sherman KE (eds) Viral insecticides for biological control. Academic, New York, pp 489–542

Vlak JM, Klinkenberg FA, Zaal KJM, Usmany M, Klinge-Roode EC, Geervliet JBF, Roosien J, van Lent JWM (1988) Functional studies on the p10 gene of *Autographa californica* nuclear polyhedrosis virus using a recombinant expressing a p10-beta-galactosidase fusion gene. J Gen Virol 69: 765–776

Vlak JM, Schouten A, Usmany M, Belsham GJ, Klinge-Roode EC, Maule AJ, Van Lent JWM, Zuidema D (1990) Expression of cauliflower mosaic virus gene 1 using a baculovirus vector based upon the p10 gene and a novel selection method. Virology 179: 312–320

Volkman LE, Goldsmith PA, Hess RT, Faulkner P (1984) Neutralization of budded *Autographa californica* nuclear polyhedrosis virus by a monoclonal antibody: identification of the target antigen. Virology 133: 354–362

Wang X, Ooi BG, Miller LK (1991) Baculovirus vectors for multiple gene expression and for occluded virus production. Gene 100: 131–137

Wathen MW, Brideau RJ, Thomsen DR (1989a) Immunization of cotton rats with the human respiratory syncytial virus F glycoprotein produced using a baculovirus vector. J Infect Dis 159: 255–264

Wathen MW, Brideau RJ, Thomsen DR, Murphy BR (1989b) Characterization of a novel human respiratory syncytial virus chimeric FG glycoprotein expressed using a baculovirus vector. J Gen Virol 70: 2625–2635

Wathen MW, Aeed PA, Elhammer AP (1991) Characterization of oligosaccharied structures on a chimeric respiratory syncytial virus protein expressed in insect cell line Sf9. Biochemistry 30: 2863–2868

Watson CJ, Hay RT (1990) Expression of adenovirus type 2 DNA polymerase in insect cells infected with a recombinant baculovirus. Nucleic Acids Res 18: 1167–1173

Webb NR, Madoulet C, Tosi P-F, Broussard DR, Sneed L, Nicolau Cl, Summers MD (1989) Cell–surface expression and purification of human CD4 produced in baculovirus-infected insect cells. Proc Natl Acad Sci USA 86: 7731–7735

Wedegaertner PB, Gill GN (1989) Activation of the purified protein tyrosine kinase domain of the epidermal growth factor receptor. J Biol Chem 264: 11346–11353

Weiss SA, Smith GC, Kalter SS, Vaughn JL (1981) Improved method for the production of insect cell cultures in volume. In Vitro 17: 495–502

Weiss SA, Belisle BW, DeGiovanni A, Godwin G, Kohler J, Summers MD (1989) Insect cells as substrates for biologicals. Dev Biol Stand 70: 271–279

Welch S-KW, Crawford SE, Estes MK (1989) Rotavirus SA11 genome segment 11 protein is a nonstructural phosphoprotein. J Virol 63: 3974–3982

Weyer U, Knight S, Possee RD (1990) Analysis of very late gene expression by *Autographa californica* nuclear polyhedrosis virus and the further development of multiple expression vectors. J Gen Virol 71: 1525–1534

Whitefleet-Smith J, Rosen E, McLinden J, Ploplis VA, Fraser MJ, Tomlinson JE, McLean JW, Castellino FJ (1989) Expression of human plasminogen cDNA in a baculovirus vector-infected insect cell system. Arch Biochem Biophys 271: 390–399

Whitt MA, Manning JS (1988) A phosphorylated 34-kDa protein and a subpopulation of polyhedrin are thiol linked to the carbohydrate layer surrounding a baculovirus occlusion body. Virology 163: 33–42

Williams GV, Rohel DZ, Kuzio J, Faulkner P (1989) A cytopathological investigation of *Autographa californica* nuclear polyhedrosis p10 gene function using insertion/deletion mutations. J Gen Virol 70: 187–202

Wilson ME, Mainprize TH, Friesen PD, Miller LK (1987) Location transcription and sequence of a baculovirus gene encoding a small arginine-rich polypeptide. J Virol 61: 661–666

Wojchowski DM, Orkin SH, Sytkowski AJ (1987) Active human erythropoietin expressed in insect cells using a baculovirus vector: a role for N-linked oligosaccharide. Biochim Biophys Acta 910: 224–232

Wood HA (1980) *Autographa californica* nuclear polyhedrosis virus-induced proteins in tissue culture. Virology 102: 21–27

Yoo D, Parker MD, Babiuk LA (1990) Analysis of the S spike (peplomer) glycoprotein of bovine coronavirus synthesized in insect cells. Virology 179: 121–128

Zhang YM, Hayes EP, McCarty TC, Dubois DK, Summers PL, Eckels KH, Chanock RM, Lai CJ (1988) Immunization of mice with dengue structural proteins and nonstructural protein NS1 expressed by a baculovirus recombinant iduces resistance to dengue virus encephalitis. J Virol 62: 3027–3031

Zuidema D, Schouten A, Usmany M, Maule AJ, Belsham GJ, Roosien J, Klinge-Roode EC, van Lent JWM, Vlak JM (1990) Expression of cauliflower mosaic virus gene I in insect cells using a novel polyhedrin-based baculovirus expression vector. J Gen Virol 71: 2201–2209

Subject Index

Current Topics in Microbiology and Immunology

Volumes published since 1981 (and still available)

Vol. 94/95: 1981. 46 figs. IV, 308 pp. ISBN 3-540-10803-3

Vol. 100: **Boehmer, H. V.; Haas, W.; Köhler, G.; Melchers, F.; Zeuthen, J. (Ed.):** T Cell Hybridomas. A Workshop at the Basel Institute for Immunology. With the collaboration Buser-Boyd, S. 1982. 52 figs. XI, 262 pp. ISBN 3-540-11535-8

Vol. 101: **Graf, Thomas; Jaenisch, Rudolf (Ed.):** Tumorviruses, Neoplastic Transformation, and Differentiation. 1982. 27 figs. VIII, 198 pp. ISBN 3-540-11665-6

Vol. 106: **Vogt, Peter K.; Koprowski, Hilary (Ed.):** Mouse Mammary Tumor Virus. 1983. 12 figs. VII, 103 pp. ISBN 3-540-12828-X

Vol. 107: **Vogt, Peter K.; Koprowski; Hilary (Ed.):** Retroviruses 2. 1983. 26 figs. VII, 180 pp. ISBN 3-540-12384-9

Vol. 109: **Doerfler, Walter (Ed.):** The Molecular Biology of Adenoviruses 1. 30 Years of Adenovirus Research 1953–1983. 1983. 69 figs. XII, 232 pp. ISBN 3-540-13034-9

Vol. 110: **Doerfler, Walter (Ed.):** The Molecular Biology of Adenoviruses 2. 30 Years of Adenovirus Research 1953–1983. 1984. 49 figs. VIII, 265 pp. ISBN 3-540-13127-2

Vol. 112: **Vogt, Peter K.; Koprowski, Hilary (Ed.):** Retroviruses 3. 1984. 19 figs. VII, 115 pp. ISBN 3-540-13307-0

Vol. 113: **Potter, Michael; Melchers, Fritz; Weigert, Martin (Ed.):** Oncogenes in B-Cell Neoplasia. Workshop at the National Cancer Institute, National Institutes of Health, Bethesda, MD March 5–7, 1984. 1984. 65 figs. XIII, 268 pp. ISBN 3-540-13597-9

Vol. 115: **Vogt, Peter K. (Ed.):** Human T-Cell Leukemia Virus. 1985. 74 figs. IX, 266 pp. ISBN 3-540-13963-X

Vol. 116: **Willis, Dawn B. (Ed.):** Iridoviridae. 1985. 65 figs. X, 173 pp. ISBN 3-540-15172-9

Vol. 122: **Potter, Michael (Ed.):** The BALB/c Mouse. Genetics and Immunology. 1985. 85 figs. XVI, 254 pp. ISBN 3-540-15834-0

Vol. 124: **Briles, David E. (Ed.):** Genetic Control of the Susceptibility to Bacterial Infection. 1986. 19 figs. XII, 175 pp. ISBN 3-540-16238-0

Vol. 125: **Wu, Henry C.; Tai, Phang C. (Ed.):** Protein Secretion and Export in Bacteria. 1986. 34 figs. X, 211 pp. ISBN 3-540-16593-2

Vol. 126: **Fleischer, Bernhard; Reimann, Jörg; Wagner, Hermann (Ed.):** Specificity and Function of Clonally Developing T-Cells. 1986. 60 figs. XV, 316 pp. ISBN 3-540-16501-0

Vol. 127: **Potter, Michael; Nadeau, Joseph H.; Cancro, Michael P. (Ed.):** The Wild Mouse in Immunology. 1986. 119 figs. XVI, 395 pp. ISBN 3-540-16657-2

Vol. 128: 1986. 12 figs. VII, 122 pp. ISBN 3-540-16621-1

Vol. 129: 1986. 43 figs., VII, 215 pp. ISBN 3-540-16834-6

Vol. 130: **Koprowski, Hilary; Melchers, Fritz (Ed.):** Peptides as Immunogens. 1986. 21 figs. X, 86 pp. ISBN 3-540-16892-3

Vol. 131: **Doerfler, Walter; Böhm, Petra (Ed.):** The Molecular Biology of Baculoviruses. 1986. 44 figs. VIII, 169 pp. ISBN 3-540-17073-1

Vol. 132: **Melchers, Fritz; Potter, Michael (Ed.):** Mechanisms in B-Cell Neoplasia. Workshop at the National Cancer Institute, National Institutes of Health, Bethesda, MD, USA, March 24–26, 1986. 1986. 156 figs. XII, 374 pp. ISBN 3-540-17048-0

Vol. 133: **Oldstone, Michael B. (Ed.):** Arenaviruses. Genes, Proteins, and Expression. 1987. 39 figs. VII, 116 pp. ISBN 3-540-17246-7

Vol. 134: **Oldstone, Michael B. (Ed.):** Arenaviruses. Biology and Immunotherapy. 1987. 33 figs. VII, 242 pp. ISBN 3-540-17322-6

Vol. 135: **Paige, Christopher J.; Gisler, Roland H. (Ed.):** Differentiation of B Lymphocytes. 1987. 25 figs. IX, 150 pp. ISBN 3-540-17470-2

Vol. 136: **Hobom, Gerd; Rott, Rudolf (Ed.):** The Molecular Biology of Bacterial Virus Systems. 1988. 20 figs. VII, 90 pp. ISBN 3-540-18513-5

Vol. 137: **Mock, Beverly; Potter, Michael (Ed.):** Genetics of Immunological Diseases. 1988. 88 figs. XI, 335 pp. ISBN 3-540-19253-0

Vol. 138: **Goebel, Werner (Ed.):** Intracellular Bacteria. 1988. 18 figs. IX, 179 pp. ISBN 3-540-50001-4

Vol. 139: **Clarke, Adrienne E.; Wilson, Ian A. (Ed.):** Carbohydrate-Protein Interaction. 1988. 35 figs. IX, 152 pp. ISBN 3-540-19378-2

Vol. 140: **Podack, Eckhard R. (Ed.):** Cytotoxic Effector Mechanisms. 1989. 24 figs. VIII, 126 pp. ISBN 3-540-50057-X

Vol. 141: **Potter, Michael; Melchers, Fritz (Ed.):** Mechanisms in B-Cell Neoplasia 1988. Workshop at the National Cancer Institute, National Institutes of Health, Bethesda, MD, USA, March 23–25, 1988. 1988. 122 figs. XIV, 340 pp. ISBN 3-540-50212-2

Vol. 142: **Schüpbach, Jörg:** Human Retrovirology. Facts and Concepts. 1989. 24 figs. 115 pp. ISBN 3-540-50455-9

Vol. 143: **Haase, Ashley, T.; Oldstone, Michael B. A. (Ed.):** In Situ Hybridization. 1989. 33 figs. XII, 90 pp. ISBN 3-540-50761-2

Vol. 144: **Knippers, Rolf; Levine, A. J. (Ed.):** Transforming Proteins of DNA Tumor Viruses. 1989. 85 figs. XIV, 300 pp. ISBN 3-540-50909-7

Vol. 145: **Oldstone, Michael B. A. (Ed.):** Molecular Mimicry. Cross-Reactivity between Microbes and Host Proteins as a Cause of Autoimmunity. 1989. 28 figs. VII, 141 pp. ISBN 3-540-50929-1

Vol. 146: **Mestecky, Jiri; McGhee, Jerry (Ed.):** New Strategies for Oral Immunization. International Symposium at the University of Alabama at Birmingham and Molecular Engineering Associates, Inc. Birmingham, AL, USA, March 21–22, 1988. 1989. 22 figs. IX, 237 pp. ISBN 3-540-50841-4

Vol. 147: **Vogt, Peter K. (Ed.):** Oncogenes. Selected Reviews. 1989. 8 figs. VII, 172 pp. ISBN 3-540-51050-8

Vol. 148: **Vogt, Peter K. (Ed.):** Oncogenes and Retroviruses. Selected Reviews. 1989. XII, 134 pp. ISBN 3-540-51051-6

Vol. 149: **Shen-Ong, Grace L. C.; Potter, Michael; Copeland, Neal G. (Ed.):** Mechanisms in Myeloid Tumorigenesis. Workshop at the National Cancer Institute, National Institutes of Health, Bethesda, MD, USA, March 22, 1988. 1989. 42 figs. X, 172 pp. ISBN 3-540-50968-2

Vol. 150: **Jann, Klaus; Jann, Barbara (Ed.):** Bacterial Capsules. 1989. 33 figs. XII, 176 pp. ISBN 3-540-51049-4

Vol. 151: **Jann, Klaus; Jann, Barbara (Ed.):** Bacterial Adhesins. 1990. 23 figs. XII, 192 pp. ISBN 3-540-51052-4

Vol. 152: **Bosma, Melvin J.; Phillips, Robert A.; Schuler, Walter (Ed.):** The Scid Mouse. Characterization and Potential Uses. EMBO Workshop held at the Basel Institute for Immunology, Basel, Switzerland, February 20–22, 1989. 1989. 72 figs. XII, 263 pp. ISBN 3-540-51512-7

Vol. 153: **Lambris, John D. (Ed.):** The Third Component of Complement. Chemistry and Biology. 1989. 38 figs. X, 251 pp. ISBN 3-540-51513-5

Vol. 154: **McDougall, James K. (Ed.):** Cytomegaloviruses. 1990. 58 figs. IX, 286 pp. ISBN 3-540-51514-3

Vol. 155: **Kaufmann, Stefan H. E. (Ed.):** T-Cell Paradigms in Parasitic and Bacterial Infections. 1990. 24 figs. IX, 162 pp. ISBN 3-540-51515-1

Vol. 156: **Dyrberg, T. (Ed.):** The Role of Viruses and the Immune System in Diabetes Mellitus. 1990. 15 figs. XI, 142 pp. ISBN 3-540-51918-1

Vol 157: **Swanstrom, Ronald; Vogt, Peter K. (Ed.):** Retroviruses. Strategies of Replication. 1990. 40 figs. XII, 260 pp. ISBN 3-540-51895-9

MIX
Papier aus verantwortungsvollen Quellen
Paper from responsible sources
FSC® C105338

If you have any concerns about our products,
you can contact us on
ProductSafety@springernature.com

In case Publisher is established outside the EU,
the EU authorized representative is:
Springer Nature Customer Service Center GmbH
Europaplatz 3, 69115 Heidelberg, Germany

Printed by Libri Plureos GmbH
in Hamburg, Germany